STUDIES IN SOVIET HISTORY AND SOCIETY

edited by Joseph S. Berliner, Seweryn Bialer,
and Sheila Fitzpatrick

Soviet Industry from Stalin to Gorbachev: Studies in
   Management and Innovation
   by Joseph S. Berliner
Research Guide to the Russian and Soviet Censuses
   edited by Ralph S. Clem
The Party and Agricultural Crisis Management in the USSR
   by Cynthia S. Kaplan
Will the Non-Russians Rebel? State, Ethnicity, and Stability in the USSR
   by Alexander J. Motyl
Revolution on the Volga: 1917 in Saratov
   by Donald J. Raleigh
Transition to Technocracy: The Structural Origins of
   the Soviet Administrative State
   by Don K. Rowney
Soviet Steel: The Challenge of Industrial Modernization in the USSR
   by Boris Z. Rumer
Revolution and Culture: The Bogdanov-Lenin Controversy
   by Zenovia A. Sochor

TRANSITION TO TECHNOCRACY

The Structural Origins of the Soviet Administrative State

DON K. ROWNEY

CORNELL UNIVERSITY PRESS

Ithaca and London

Copyright © 1989 by Don K. Rowney

All rights reserved. Except for brief quotations in a review, this book, or parts thereof, must not be reproduced in any form without permission in writing from the publisher. For information, address Cornell University Press, 124 Roberts Place, Ithaca, New York 14850.

First published 1989 by Cornell University Press.

International Standard Book Number 0-8014-2183-7
Library of Congress Catalog Card Number 88-47925

Printed in the United States of America

Librarians: Library of Congress cataloging information appears on the last page of the book.

The paper in this book is acid-free and meets the guidelines for permanence and durability of the Committee on Production Guidelines for Book Longevity of the Council on Library Resources.

For Susan
With Love and Hope

Inside the stuffy coach, seated
Among the plain and lowliest,
I fear I yield myself to feelings
I sucked in at my mother's breast.

But, brooding over past reverses
And years of our penury and war,
In silence I discern my people's
Incomparable traits once more.

And worshipful, I humbly watch
Old peasant women, Muscovites,
Plain artisans, plain laborers,
Young students, and suburbanites.

I see no traces of subjection
Born of unhappiness, dismay,
Or want. They bear their daily trials
Like masters who have come to stay.

Disposed in every sort of posture,
In little knots, in quiet nooks,
The children and the young sit still,
Engrossed, like experts, reading books.

Then Moscow greets us in a mist
Of darkness turning silver grey
When, leaving the underground station,
We come into the light of day.

And crowding to the exits, going
Their way, our youth and future spread
The freshness of wildcherry soap
And the smell of honeyed gingerbread.

BORIS PASTERNAK, *On Early Trains*, 1941

Contents

	Figures and Tables	ix
	Preface	xi
1	Introduction: Revolution, Bureaucracy, and Technocracy in Russia	1
2	Domestic Administration in Structural Perspective, before 1917	19
3	The First Structural Transformation, 1917–1918	65
4	The Transformation of Personnel in Central Government, 1917–1923	94
5	The Communist Party and the Soviet Administrative Elite, 1922–1930	124
6	Bureaucratic Structure and Revolution: Elite Survival and Replacement	175
	Bibliography	211
	Index	233

Figures and Tables

FIGURES

1. Distribution of provincial officials "predicted" from population and mortality rate, c. 1860   53
2. Distribution of provincial officials "predicted" from urban and manufacturing population, c. 1897   54
3. Structure of the Commissariat of Health, 1922   90
4. Social background and employment status of Party members, 1925   149
5. Social background and employment status of Party members, 1926   150
6. Social background and employment status of Party members, 1927   151
7. Social background and employment status of Party members, 1928   152

TABLES

1. Change in landholding among officials   36
2. Ministry of Internal Affairs: Social origins of high officials, 1905   42
3. Ministry of Internal Affairs: Social origins of high officials, 1916   42
4. Prior positions of provincial governors   44
5. Prior positions of provincial vice-governors   45
6. Sources of land held by imperial civil servants   48
7. Distribution of officials correlated with demographic data, 1860   51
8. Distribution of officials correlated with demographic data, 1897   52
9. Population characteristics of selected provinces   59

10. Patterns of peasant action, 1917 71
11. Aggregate and per capita incidents of peasant action 74
12. Status of health/medical administrators in central agencies, 1912 100
13. Professional roles in 1912 of Health Commissariat physicians 105
14. High commissariat officials employed by the state before 1917 107
15. Prerevolutionary records of holdovers 109
16. Appointment date of Posts and Telegraphs employees, 1927 113
17. Appointment date of Posts and Telegraphs administrative personnel, 1927 113
18. Career activities before and after 1917 of Moscow officials, 1922 115
19. Communists among officials, 1922 and 1929 132
20. Age distribution of Moscow officials, 1922 134
21. Age distribution of All Union and Union Republic officials, 1929 134
22. Education of Moscow officials, 1922 135
23. Education of All Union officials, 1929 136
24. Education of Union Republic officials, 1929 137
25. Education of Union and Union Republic officials, 1927 and 1929 144
26. Education of officials in regional administrations, 1927 and 1929 145
27. Education of officials in local administrations, 1927 and 1929 145
28. Social origins of Communists in state and white-collar roles 153
29. Employment status of class categories of Communists 155
30. Background of Party members in white-collar positions 157
31. Commissarial staff distribution by office level and class 158
32. Communist and non-Communist access to white-collar status 161
33. Communists in administrative positions, 1927 162
34. Skill level of worker-Communists correlated with socioeconomic characteristics 167
35. Mobility patterns of Communists in rural areas 168

Preface

This book began in my desire to evaluate the proposition that the 1920s witnessed the birth of a Soviet state technocracy. I was interested to learn whether we could discern, at the time of Lenin's death, for example, the outlines of an administrative system that displayed certain characteristics normally thought to be incompatible with the capricious politics of autocracy and dictatorship. In particular I wanted to know whether individuals professionally trained in technical and scientific fields had *both* important operational roles *and* the formal authority and power to allocate resources such as money and promotions from inside their various state administrative offices.

In principle, I had good reasons to think that throughout the modern history of Russia a transition to technocracy—that "happy" era, as Lenin put it, when engineers and agronomists would do most of the talking—was in process despite the tendency of politicians to exercise enormous political power arbitrarily. For one thing, Lenin himself often supported the idea that technicians and specialists, rather than politicians, should be in charge. For another, my previous research on the late nineteenth-century Imperial bureaucracy revealed that some government agencies were then moving markedly toward a technocratic mode of operation. And, although it would be a very long time in coming, it now seems evident that Mikhail S. Gorbachev's *perestroika* is the fruition of a historical process of administrative development. Gorbachev's policies have already opened new fields of influence and operation to many segments of Soviet society. Among these, certainly, is an administrative core consisting of a highly professionalized and specialized intelligentsia. Though they maintain some distance from the arbitrary and centralizing political traditions of the past half century, these officials are easily *dirigiste* enough to be described as technocratic.

My final and most important reason for thinking of the 1920s as an era of administrative transition emerged early in my research. It quickly became clear that if we look at the details of the structure and development of Soviet administration after the Revolution of 1917, we find technically shaped administrative offices growing rapidly and the numbers of technically trained administrators increasing. This phenomenon was of course important in itself for administrative change in Russia. But it had social implications as well. Because of the leviathan size and power of bureaucracy in Russia, the postrevolutionary transformation of *offices* inevitably produced a transformation of *officials* on a scale previously unknown anywhere in Europe. These administrative and social changes alone pointed me toward the need to explain both the origins of the Soviet state technocracy and the historical interaction between this technocracy and Russian society.

My research here is thus concerned principally with what I am calling the transitional generation of administrators in Russia and the USSR, that body of officialdom that crossed the revolutionary divide of 1917 and melded into the first generation of truly Soviet administrators. I am interested in these officials as such but also because they were witnesses to and participants in the process that transformed both the Russian bureaucracy and huge segments of Russian society.

Two of the most important arguments against the proposition that either the pre- or postrevolutionary eras in Russian or Soviet history saw the birth of technocracy are based on the tradition of arbitrary political intervention and the function of ideology as a source of countervailing values in administration. Each of these concepts has served as the thematic focus of countless histories of the Russian and Soviet states and their administrations. In the view of some, the importance of arbitrary and unpredictable political intervention in administrative affairs in the tsarist era is exceeded only by that of the raw and annihilating power of totalitarian political dictatorship in the Soviet era. The implied or expressed conclusion has inevitably been that such all-encompassing, arbitrary politics and their accompanying ideologies render even a legalistic, Weberian bureaucracy impossible. A true technocracy, in which disinterested expertise and rationality combine with a legally secure bureaucracy, has thus seemed unattainable for Russia and the USSR.

Preface xiii

In my view, the repeated episodes of tsarist and Soviet political intervention in civil administrative affairs are significant in the broad, long-run sense that politics is important in every kind of organization. The control of human organizations always requires power. I think the crucial issue for Soviet history is whether political intervention can be shown to work consistently and effectively against the influence of technocrats in decisions important to the organization, such as hiring, firing, and the allocation of promotions and internal resources. One can easily show that political intervention does deflect the impact of technocrats in specific administrative decisions and selected examples of other bureaucratic behavior, but there is little evidence, in the history of either Russia or the USSR, that such intervening was systematically done over considerable periods of time for the bulk of the civil administration. Thus, in this book, I focus principally not on the exceptional episodes of political intervention but on the typical or normal office and official as I try to characterize patterns of change across time.

Ideology, specifically Marxist-Leninist ideology on the one hand and tsarist-Russian cultural values on the other, has certainly played a powerful countervailing role in the development of Russian and Soviet technocratic behavior. Since this statement is equally true for the twentieth century and for the nineteenth, I look carefully at the historical background, or patrimony of social and political behavior, which the state administration confronted in the revolutionary era.

Briefly, the evidence that the 1920s saw the nascence of a Soviet technocracy can be divided into three major categories: (1) that concerned with the structural transformation from generalist to specialist bureaucracy, which began as early as the era of Alexander I and proceeded rapidly after 1917; (2) that showing the inexorably increasing numbers and progressively elevated roles of specialists, a notable process in the second half of the nineteenth century, which took off rapidly after 1917; (3) that of the ever-increasing bureaucratization which, in Russia, was an inevitable consequence of the extension of government further and further into the details of social organization.

None of these fields of evidence alone suffices to sustain the conclusion that the post-1917 era saw the beginning of Soviet technocracy. Indeed, even taken together they do not furnish us with enough information to evaluate the process, because such major factors as the social and political roles of the nobility, the peasantry, and the Communist Party must also be studied. What we do find, however, is that a consideration of the complex interaction of these factors across the historical landscape of nineteenth- and twentieth-century Russia enables us to understand what technocracy means in the modern USSR, just as it equips us better to understand the social implications of administrative change and the structural thrust of state administration into the future.

In part because this book is fundamentally a large-scale social history and in part because of the nature of the information upon which it is based, many of the arguments presented here will rely on numbers—of people and of events—as well as on relative scales and orders of events. Although much of the analysis is therefore quantitative, I have attempted throughout to use techniques that are readily understood by the reader unversed in such analysis.

The list of institutions and persons to whom I am indebted for help in the preparation of this book is a long one. I am constrained, therefore, to acknowledge individually only a few of them, either for direct assistance in development of the research or, often, for unflagging interest and support of a long and sometimes frustrating project.

Among those who read and commented on the manuscript, I am especially grateful to Susan R. Carlton, Robert W. Davies, J. Arch Getty, John J. Markovic, Walter M. Pintner, and William G. Rosenberg. For help in the preparation of the text, I offer a special note of thanks to Anisa Miller-Pogacar, my colleague Bernard Sternsher, and the staff of Cornell University Press—especially my copy editor, Patricia Sterling, and my editor, John G. Ackerman.

Among those who provided general comments, suggestions, and sometimes badly needed moral support, I include John A. Armstrong, Jean Bonamour, Adam Bromke, William J. Hudson, and Michel Lesage together with his colleagues at the National Center for Scientific Research in Ivry, France.

Soviet colleagues to whom I am particularly indebted include V. Z. Drobizhev, Ivan D. Koval'chenko, Sergei Miakushev, and Iurii A. Poliakov and his associates at the Institute of History of the Academy of Sciences of the USSR.

American professional librarians are generally so competent and helpful that it is impossible to single out any for particular thanks and praise in spite of the fact that I relied upon many of them again and again. In Moscow, I benefited from the professional standards of the staff at the Institute for Scientific Study in the Social Sciences. In Paris, I was fortunate to be able to rely for assistance on the excellent staff at the Library for International Contemporary Documentation.

Obviously a project of this kind would be impossible without generous support for research from public and private foundations. In this case, while it is important to note that my findings do not necessarily represent their views, I am happy to acknowledge the assistance of the Academy of Sciences of the USSR, the American Council of Learned

Societies, the International Research and Exchanges Board, the National Endowment for the Humanities, and, last but certainly not least, the Faculty Research Committee and the Department of History of Bowling Green State University.

The lines from Boris Pasternak's "On Early Trains" are quoted from Pasternak's *Poems*, trans. Eugene M. Kayden (Ann Arbor: University of Michigan Press, 1959). I am grateful for permission to reprint them.

DON K. ROWNEY

Toledo, Ohio

TRANSITION TO TECHNOCRACY

CHAPTER 1

Introduction: Revolution, Bureaucracy, and Technocracy in Russia

I think we are now at the most fundamental turning point which, from every perspective, will mark the beginning of major successes for Soviet power. Henceforth, at the tribune of All-Russian meetings not only will politicians and administrators hold forth, but also engineers and agronomists. This is the beginning of a very happy era, when politicians will grow ever fewer in number, when people will speak of politics more rarely and at less length, and when engineers and agronomists will do most of the talking.

V. I. Lenin

Moreover, instead of general knowledge, every man should be required to know only that which is necessary for the service to which he wants to devote himself.

N. M. KARAMZIN

The two statements that serve as epigraphs above[1] were separated by more than one hundred turbulent years as measured by the *Kremlevskie kuranty*, the great chiming clocks in the Kremlin's Spasskii Tower. Indeed, the differences in social and political perspectives between these authors were even greater than a century of extraordinary polit-

---

1. V. I. Lenin, *Polnoe sobranie sochinenii* (hereafter PSS), 5th ed. (Moscow, 1958–65), 42:156; N. M. Karamzin, in Richard Pipes, ed., *Karamzin's Memoir on Ancient and Modern Russia* (Cambridge, Mass., 1966), 160–61. (All translations not indicated otherwise are my own.) Karamzin's scorn was roused by the Civil Service Examination Act of 1809 (*Polnoe sobranie zakonov* [hereafter PSZ], 1809, no. 23711), the work of Mikhail

ical, social, and economic change might first suggest. Lenin was the acknowledged leader of the most dramatic political upheaval of the twentieth century. Nicholas Karamzin, for all his daring as a rhetorician, was as reactionary as any major historian Russia would produce. Yet both statements bespeak a shared belief in the adequacy and importance of the specialist in state administration. This faith has been a wellspring of both criticism and hope for students and practitioners of Russian administration since the eighteenth century, and the enduring significance in Russia of administrative specialization and of its modern variant, technocracy, serves as my central theme.

But this book is not a history of Russian and Soviet technocracy or a collective biography of technocrats. First, my focus is limited to the emergence of the expert within one important group of institutions—the civil bureaucracies of Imperial Russia and the U.S.S.R. This focus has important implications. It leaves out of consideration the emergence of technical expertise in such spheres as private business and the professions in Imperial Russia. More significantly, however, such a focus concentrates attention on major *state* institutions in a society where, in modern times, the state has always been enormously powerful.

Second, my interest in the emergence of experts is but a part—the most important part, to be sure—of a broader interest in the social and organizational structures composing the Imperial and Soviet bureaucracies. The emergence of technocracy is examined, and better understood, in the context of both social and organizational change in official Russia. Both the structural evolution of civil administration and the social histories of state servitors, landed and landless nobility, nonnoble officials, workers, and peasants take their places at various stages in the development of this analysis.

In certain respects, then, this study is broad and complex. The role of civil administration has been so intricately combined with the evolution of Russian society that it is impossible to study any important aspect of Russian civil administration without writing about segments of Russian society. Not only has civil administration extended its functions into the most intimate details of the lives of all ordinary people in Russia, but it has become a major source or framework for the formulation of economic and educational policy. Not only has it played a major part in shaping economic policy, but it has become the major

---

Speranskii. Pipes (237) says that Karamzin "misconstrued the intent of the law." That may be; but given Karamzin's outlook on both politics and autocracy, as explained by Pipes (48–49, 62–63), one must assume that there is more here than a simple misunderstanding. Officials, Karamzin is saying, have specific and important jobs to do; irrelevant education only interferes and, we might interpolate, tempts them into dangerous activities—corruption and plotting.

employer in Soviet society. Not only has it served as nearly the only link between public and private life, but it has become the social benefactor and legal guardian of every citizen.

I do not take the position that the technocratic transformation of Russian administration is a completed process. The major personnel and structural changes that were inaugurated in the USSR in 1985 have been described as technocratic.[2] It is not assumed here, however, that they will be entirely successful any time soon.

Certainly, as measured by the high standards used by some critics of modern Russian administration, the happy era envisioned by Lenin is still far away. But much of the discourse about technocracy concentrates on the question of control; in societies—whether democratic or dictatorial—where expertise is crucial, the question of whether experts control or are controlled is chronic. Consequently, the encroachment of experts into institutions from which they have previously been excluded always raises political questions. As broader areas of organizational function fall under the authority of experts, the question inevitably arises, who is really in charge here? The people? the monarch? the ruling party? the dictator?

But the Russian technocratic transformation continues. As Soviet historian Roy Medvedev puts it: "Life is becoming more and more complex and bureaucrats are constantly pursued by failure. This is the reason why scientists and specialists are being drawn into the *apparat*; committees and commissions of experts are being created, but at the same time undemocratic methods of administration remain intact. In other words, the bureaucrats are being replaced by knowledgeable and more efficient technocrats. Very possibly the major development of the next ten or fifteen years will be the transition from bureaucracy to technocracy.[3]

Like Lenin, it is interesting to note, Medvedev sees technocracy both as something positive, a moving toward competent, disinterested, technical control, and also as something negative, a moving away—for Lenin, from politics; for Medvedev, from bureaucracy. These expectations inspire two questions. *Does* the ascendance of the technocrat necessarily imply the eclipse of the "bureaucrat" as Medvedev uses the term? *Does* it mean the end of politics? The history of the extension of technical administrative roles in Russia, as we shall see, has not necessarily implied either depoliticization or debureaucratization. The process of drawing scientists and specialists into the *apparat* in Russia

---

2. Seweryn Bialer and Joan Afferica, "Gorbachev's Preference for Technocrats," *New York Times*, February 11, 1986; see also by the same authors "The Genesis of Gorbachev's World," *Foreign Affairs* 64, no. 3 (1985), 605–44.
3. Roy Medvedev, *On Socialist Democracy* (New York, 1975), 300.

has been in train for a long time. It has evolved through several complex stages in its impact on both the state administration and segments of Russian-Soviet society. This book attempts not only to trace specific aspects of the process of technocratic development but also to describe the consequences of its involvement with bureaucracy and society.

Undoubtedly, this is an ambitious enterprise. Indeed, it is legitimate to ask whether it is reasonable to speak of technocracy in a society where arbitrary political behavior has been the historical norm. Technocracy, after all, requires more than the application of scientific and technical information to organizational problems. It also implies the exercise of impersonal, objective judgment by decision-makers—an image that is far from our historical impressions of Nicholas II or Stalin.

By the same token, one may ask whether there can be any substance to a developing or maturing relationship between bureaucracy and technocracy when the two concepts, as Medvedev's comment suggests, are often regarded as inimical. It is just as important, moreover, to ask how we can talk of a "transition to technocracy" in a country where industrialization, in its contemporary sense, was barely one generation old. By world standards, Russian technology—and therefore its manipulation—was primitive at the time of the Russian Revolution of 1917.

In fact, we shall see that these elements—bureaucratic administration, technical control of bureaucracy and of society through bureaucracy, and political intrusion into bureaucracy—coexisted not only during the generation that followed the Revolution of 1917 but also during the generation that preceded it. That their interrelation was sporadic and not always predictable, while consistently admitting of a greater role for specialization, is the circumstance that permits us to speak of a "transition" to technocracy. It is my position that such a transitional period straddled the Revolution of 1917, including the last prerevolutionary generation and the first postrevolutionary one. By the end of the 1920s—not, it should be stressed, the 1930s—the transitional era was substantially complete: the social and organizational foundations for the state edifice that was eventually built in the Stalin period were finished, and these foundations required not merely an apparatus staffed by a "technical intelligentsia" but a technocratic administrative apparatus.

Up to a point, then, this book adopts the view of E. H. Carr and R. W. Davies that the edifice of the Stalinist, Five Year Plan administrations was already well under construction before Stalin himself was fully entrenched in power.[4] The data, however, oblige one to go further.

---

4. E. H. Carr and R. W. Davies, *Foundations of A Planned Economy, 1926–1929.* 2 vols. (London, 1969–71).

Because of the place of civil administration in tsarist Russia and because of its complicated interrelations with Russian society, it is both fair and essential to say that the foundations were truly laid at least as early as the 1890s and in the first postrevolutionary years.

Definitions: Technocracy

In order to be clear about the principal terms and concepts used here, let me define them explicitly. First, technocracy. By "technocracy" I mean organizations in which individuals trained or professionally experienced in modern technical fields (for example, the sciences, engineering, medicine) have *both* dominant operational roles *and* the formal authority and power to allocate resources such as money and promotions and to decide policy issues that are relevant to the organization.

This means, for instance, that in a technocracy civil engineers are in charge of transport ministries; physicians of health ministries; communications engineers of ministries of telephone and telegraph. Specialists—not administrative generalists who skip from one ministry to another—hire and fire; they allocate the budget; they decide what needs to be done first and what can wait until later. It *may* also mean that in competition for the top administrative posts, persons with the right technical credentials and experience will succeed more often than those without those credentials, regardless of personal wealth, family background, or "connections." But in a society like Russia, where formal education often has had a social status bias, the connection between career and social status is complex.

The reader should note that, owing to my specific focus on civil bureaucracy, this definition is narrower than those often used in both the historical and sociological literature on technocracy. For example, Kendall Bailes uses the term more broadly. I take the following statement to mean that technocracy is achieved when experts displace generalist politicians throughout society and not merely in specific organizations, however important they may be: "The term 'technocratic trend' is used here ... to indicate any movement among technical specialists that urges them to develop a wider sense of social responsibility for the use of their technical knowledge, and particularly urges them to take an important role in policy formation.... engineers should not simply be content to be the technical executors of other men's policies, but should become politicians themselves."[5]

Of course, there are questions of degree. For example, how many

---

5. Kendall E. Bailes, "The Politics of Technology: Stalin and Technocratic Thinking among Soviet Engineers," *American Historical Review* 79, no. 2 (1974), 449.

organizational positions or how many entire organizations have to fall under specialists' control in order to have a true technocracy? It is precisely the fact that technocratic control is something that develops over a period of historical time that underscores the importance of the degree of technocracy and allows us to see that there will be such a thing as a *transition* to technocracy, a period during which offices or bureaucratic functions that demand technocratic roles fall increasingly under specialists' control. In Russia, in particular, administrative activities and legal forms were changing in a direction that demanded increasingly important roles for specialists long before specialists began to play *controlling* roles in their native administrations, let alone in the society at large.

The question of what remains of nontechnical political control in technocracy has exercised students of the history of both authoritarian and postauthoritarian societies.[6] It has been of equal concern to students of politics and technical elites in the developed democracies where the expert view on everything from the waging of war to sex education has systematically eroded the perceived range of democratic choice and individual responsibility.[7]

The use of such terms as "scientific intelligentsia" or "technical intelligentsia" as substitutes for "technocrats" may sometimes avoid direct consideration of the problem of control. Alternatively, this usage assumes that the problem has already been resolved, with nonspecialist politicians firmly in charge. For example, Bailes focuses on the question of competition between technocrats and politicians at the highest political levels without, however, considering in any detail the problem of organizational or bureaucratic control. In the introduction to his

---

6. John Charles Guse, "The Spirit of Plassenburg: Technology and Ideology in the Third Reich" (Ph.D. diss., University of Nebraska, Lincoln, 1981); Albert Speer, *Infiltration* (New York, 1981); José Vincente Casanova, "The Opus Dei Ethic and the Modernization of Spain" (Ph.D. diss., New School for Social Research, 1982); William H. McNeill, *The Pursuit of Power: Technology, Armed Force, and Society since A.D. 1000* (Chicago, 1982).

7. Charles S. Maier, "Between Taylorism and Technocracy: European Ideologies and the Vision of Industrial Productivity," *Journal of Contemporary History* 5, no. 2 (1970), 27–61; P. Savigear, "Some Political Consequences of Technocracy," *Journal of European Studies* 1, no. 2 (1971), 149–60; Reynold A. Reimer, "The National School of Administration: Selection and Preparation of an Elite in Post-War France" (Ph.D. diss., Johns Hopkins University, 1977); Ezra N. Suleiman, *Politics, Power, and Bureaucracy in France* (Princeton, N.J., 1974); Robert B. Carlisle, "The Birth of Technocracy: Science, Society, and Saint-Simonians," *Journal of the History of Ideas* 35, no. 3 (1974), 445–64; John G. Gunnell, "The Technocratic Image and the Theory of Technocracy," *Technology and Culture* 23, no. 3 (1982), 392–416; Guy Alchon, "Technocratic Social Science and the Rise of Managed Capitalism, 1910–1933" (Ph.D. diss., University of Iowa, 1982). On the problem of contradictions between technocracy and politics in modern organizations, see Herbert Simon, *Administrative Behavior* (New York, 1948), 20–44.

study of the development of the technical elite in the early years of Soviet history, Bailes is at pains to explain what he means by "technical intelligentsia" and to be explicit about his model of Soviet politics, which combines features of totalitarian and group conflict theories.[8] So far as I can tell, however, throughout this excellent book Bailes assumes that politicians will be the ultimate initiators and arbiters of whatever decisions affect the fate of technical elites as a group, whether within or beyond the boundaries of the organizations to which they are attached.

In a study that focuses on industrial managerial and technical personnel in the post–1917 era, Nicholas Lampert evaluates the changing relation between this technical intelligentsia and political leadership.[9] His perspective helps us understand the degree to which conflict between industrial specialists of pre- and postrevolutionary backgrounds was at the center of political interest at the end of the 1920s, but his focus is at the level of the firm, and it too seems to give priority to politicians in questions about control.

This book adopts an approach that does not prejudge whether politicians, specialists, or nonspecialist bureaucrats are in control at any given juncture and at any given level of organizational or social authority. It shows, moreover, that at specific critical moments both before and after the Revolution of 1917, state bureaucratic elites generally and technical specialists in particular achieved objectives that were either unanticipated or directly contrary to those anticipated by politicians.

An approach that assumes political control in the development of both civil and industrial scientific administration in the 1920s and 1930s seems standard for Soviet historians.[10] But the rapid and obvious ascent to power of engineers and scientists in the post–World War II strongholds of Soviet political power—for example, the All-Union and Republican Councils of Ministers—suggests the need to reevaluate these assumptions.[11] John A. Armstrong concluded in a 1965 study of

---

8. Kendall E. Bailes, *Technology and Society under Lenin and Stalin* (Princeton, N.J., 1978), 4, 12–13.

9. Nicholas Lampert, *The Technical Intelligentsia and the Soviet State* (New York, 1979).

10. L. V. Ivanova, *Formirovaniie sovetskoi nauchnoi intelligentsii, 1917–1927* (Moscow, 1980); S. A. Fediukin, *Sovetskaia vlast' i burzhuaznye spetsialisty* (Moscow, 1972), and *Privlechenie burzhuaznoi tekhnicheskoi intelligentsii k sotsialisticheskomu stroitel'stvu v SSSR* (Moscow, 1960); D. I. Sol'skii, "NOT i voprosy deloproizvodstva (1918–1924)," *Sovetskie arkhivy* (1969), 47–52; A. I. Lutchenko, "Rukovodstvo KPSS: Formirovaniem kadrov tekhnicheskoi intelligentsii (1926–1933)," *Voprosy istorii KPSS* 2 (1966), 29–42; V. A. Ulianovskaia, *Formirovanie nauchnoi intelligentsii v SSSR, 1917–1937* (Moscow, 1966).

11. These are individuals to whom Ezra N. Suleiman refers as "secondarily specialized; that is, despite their control over a particular sector, they come to hold key positions

the "sources" of Soviet administrative behavior that "an engineering approach or technocratic approach toward human problems is more characteristic of Soviet than of Western European administrators."[12] The era when specialists of many stripes, but especially engineers, began to dominate in the political roles of the USSR is comparatively recent, which simply underscores the fact that in terms of human generations the era of the purges, the First Five Year Plan, the New Economic Policy, and even the Revolution of 1917 are all a part of our recent past, crowded together by our perspective on historical time, forcing themselves on our attention and continually shaping our present and future.

Given the scale of the "transformational" problems in administrative development, the glacial pace at which large organizations seem to change, and the degree to which such transformations become politicized, one ought to assume that the foundations for specialist dominance were laid not in the 1930s but in the 1920s, if not before. That historians fail to make such an assumption merely reflects the fact that we do not as yet have a good understanding of the administrative and social history of the Soviet state in the 1920s; that we do not as yet understand clearly what changed and what endured when Russian society passed through the crucible of revolution; and finally, that political values and the struggle of the postrevolutionary leadership and its opponents over international legitimacy and domestic control continue to exert their distracting and distorting influence on scholarly

---

in other sectors." See his *Elites in French Society: The Politics of Survival* (Princeton, N.J., 1978), 13. For major discussions and descriptions of Soviet "secondary specialists," see Jerry F. Hough, *The Soviet Prefects: The Local Party Organs in Industrial Decision-Making* (Cambridge, Mass., 1969), and *Soviet Leadership in Transition* (Washington, D.C., 1980); Grey Hodnett, *Leadership in the Soviet National Republics: A Quantitative Study of Recruitment Policy* (Oakville, Ont., 1978); Eric Vigne, "URSS: Les ingénieurs prennent le pouvoir" *Histoire* 26 (1980), 88–90; John A. Armstrong, "Sources of Administrative Behavior: Some Soviet and Western European Comparisons," *American Political Science Review* 59, no. 3 (1965), 643–655. Still, the dominance of Soviet politics by individuals who are "secondarily specialized" is not accepted by many as tantamount to technocracy in the USSR. In Mary McAuley, "The Hunting of the Hierarchy: RSFSR OBKOM First Secretaries and the Central Committee," *Soviet Studies* 26, no. 4 (1974), 473–501, the economic importance of a region and comparatively rapid development are identified as partial explanations for the Central Committee status of regional political leaders under Khrushchev and the early Brezhnev; specialization then becomes a kind of incidental consequence of modernization. For a variant of this view, see T. W. Luke and Carl Boggs, "Soviet Subimperialism and the Crisis of Bureaucratic Centralism," *Studies in Comparative Communism* 15, no. 2 (1982), 95–124: "Under bureaucratic centralism, the Soviet Union has made many advances toward becoming an *industrial* society; however, plainly it has failed to transform the country into a *technological* society." For the view, common in America, that Soviet technical administrators are merely politicians or are mainly motivated by politics, see Zbigniew K. Brzezinski and Samuel P. Huntington, *Political Power: USA/USSR* (New York, 1963), 150ff.

12. Armstrong, "Sources of Administrative Behavior," 654.

Definitions: Bureaucracy

In the context of history and of technocracy, the pejorative application of the word "bureaucracy" comes easily in the Russian usage; moreover, if we can take Medvedev's word, there is an assumption of incompatibility between bureaucracy and technocracy. Because of long-established Western academic practice, however, this book uses the term "bureaucracy" in its ostensibly nonpejorative sense of hierarchical, functionally specialized, complex organization.

Even Medvedev is not so doctrinaire that he is unwilling to allow a positive sense for the term: "The expression 'bureaucratic apparatus' sometimes refers in a general way to the entire machinery of government, and all government employees are called 'bureaucrats,' not necessarily in a pejorative sense. In this usage, 'bureaucrat' is just a synonym for 'administrator,' and bureaucracy has a positive role in the sense that no state, whether capitalist or socialist, has as yet been able to function without administrative machinery at various levels."[13]

The use of the word in an "essentially neutral" fashion offers what Roger Pethybridge, toward the end of an extended discussion of bureaucracy in the Russian context, calls the "purely methodological" advantages of permitting us "to lay stress on the historical and social reasons in Russia for the growth of Soviet administration."[14] In fact, these advantages are available through the use of no other term. Beyond that, however, the term usefully situates the organizational aspect of this research within the large body of sociological and political research published in the past hundred years. Finally, it is also clear that in this neutral sense, "bureaucracy" describes the organizational structure within which technocrats and technocratic activity are located. Technocrats, like other administrators in modern, complex organizations, are functionally specialized and subordinate to hierarchic control. The extent to which the hierarchy is composed of specialists—that is, other technocrats—speaks to the degree of technocratic extension but not to the question of whether the organization is bureaucratic.

It is helpful to an understanding of the sense in which the term "bureaucracy" is used here to recognize that organizations tend to take

---

13. Medvedev, *On Socialist Democracy*, 291.
14. Roger Pethybridge, *The Social Prelude to Stalinism* (New York, 1974), 255. For the same topic treated from a Weberian perspective, see Reinhard Bendix, *Work and Authority in Industry: Ideologies and Management in the Course of Industrialization* (New York, 1963), chaps. 3, 4.

only three alternative forms: patriarchic (or familial), collegial, and bureaucratic. For example, the patriarchy and, sometimes, the matriarchy are common in traditional societies perhaps because these are straightforward extensions of family organization, the earliest organizational experience that most human beings have.[15] But "patriarchy" also appropriately describes situations in advanced industrial societies where authority is vested in some one individual as a personality. It is not unusual to find vestiges of patriarchy in the persons of college presidents and corporate chief executives. We frequently encounter organizational circumstances contradicting Henry Maine's assertion that "it is Contract which replaces by degrees those forms of reciprocity in rights and duties which have their origin in the Family."[16] Far from "having steadily moved toward a phase of social order that is alien to Family tradition," we often find examples in modern organizational relations in which patriarchal roles are essential to organizational survival.

The college, or council, is found in traditional societies whose authority has to be collective, and information or experience needs to be pooled. Decision-making in Russian villages before collectivization in the 1930s was often collegial. Similarly, the soviet, or collegium, was commonly part of the highest level of ministry management in nineteenth-century Russia. But the council is also very common in modern industrial, scientific, and educational enterprises, in democratically controlled organizations, and generally in situations where it is necessary either to dilute or to pool individual authority or responsibility. As Bernard Silberman shows, moreover, the collegial approach to power management is useful when powerful bureaucratic or patriarchal elites confront unprecedented problems.[17]

We encounter bureaucracy, by contrast, where it is necessary to control authority by making it accountable to superior authority and where it is desirable to organize the expertise of specialized operations in a predictable manner. Bureaucracy also tends to be found where organizational tasks are so complex as to require prolonged periods of time for execution and where the necessary mixture of information and other resources exceeds the capacities of a given number of persons; for example, broad open-ended governmental programs such as law enforcement, defense, and tax gathering—as well as complex manufacturing

---

15. Claude Lévi-Strauss, *Tristes Tropiques* (New York, 1975), and *The View from Afar* (New York, 1985).
16. Henry Maine, *Ancient Law* (New York, 1931), 99.
17. Bernard Silberman, *Ministers of Modernization: Elite Mobility in the Meiji Restoration, 1868–1873* (Tucson, Ariz., 1964).

operations that are repeated indefinitely—are well suited to bureaucratic organization.

Bureaucrats' professional tasks are so defined that they will integrate readily and predictably with the tasks of many other individuals. Similarly, they are so defined as to be separable from the person who holds a given office. Official tasks and official authority are, as Max Weber pointed out, attached to the office, not to the individual: "Members of the corporate group, insofar as they obey a person in authority, do not owe this obedience to him as an individual, but to the impersonal order."[18]

It is worth noting, however, that the distinction between person and authority in bureaucracy is often blurred. Who has not encountered a clerk, secretary, or manager whose personality dominates an entire organization, subordinate and superior alike? But this pervasive circumstance merely demonstrates that the categories are not absolute. Elements of patriarchy may well be found in bureaucracy; elements of bureaucracy may well be found in what are supposed to be collegial organizations. As Reinhard Bendix notes in comparing patrimonial and bureaucratic administrations: "The benchmark concepts of social structures can encompass a range of historical experience.... To avoid the reification of the type, that is, the fallacy of attributing to a social structure a concreteness it does not possess, we must see these 'attributes' as objects of action by specific groups."[19]

In the classic bureaucracy, by contrast to other types of organizations, an advancing career is one in which authority, scope of operation, and emoluments increase from time to time. Since all of these attach in a predictable way to specific offices, the person who moves from one office to another usually expects improving prospects as a function of merit and seniority, rather than of such characteristics as family ties, wealth, or social connections. Given the power and large numbers of bureaucratic officeholders in modern organizations, the fulfillment or frustration of these career expectations can have broad social and political implications.

In Russia, bureaucratic advancement came more predictably in the prerevolutionary era than in the decade that followed. The rupture of established procedure and expectations was one of the most distinct and important consequences of the Revolution of 1917. When old bureaucratic officeholders were displaced or denied advancement, their organizational roles nevertheless had to be filled, even though the new

---

18. Max Weber, *The Theory of Social and Economic Organization* (New York, 1947), 330.

19. Reinhard Bendix, *Nation-Building and Citizenship: Studies of Our Changing Social Order*, rev. ed. (Berkeley, Calif., 1977), 137.

political leadership harbored feelings of revulsion for "bureaucratic" (now in the pejorative sense) function and status. And it was critically important both to the course of the revolution and to the further development of technocratic administration to fill these positions mainly from the lower classes, to assign to them men and women who began to enjoy the perquisites of indoor, sit-down work—"gentleman's work"—for the first time in their lives.

An important part of the argument presented here, then, is that the post–1917 bureaucracy was both old and new. On the one hand, it did not spring fully formed from the brow of the revolution; on the contrary, during the 1920s the Russian civil bureaucracy extended certain patterns of social transformation and modification of its functions that had begun in the nineteenth century. Thus, while later chapters call attention to the importance of ruptures in long-established patterns of bureaucratic control in Russia, they also disagree sharply with the view of those who see the post–1917 civil state administration as a new, purposely made tool for the attainment of legitimation and control.[20]

Nevertheless, this study also reveals that at the time of revolutionary crisis the Russian bureaucracy was an indispensable constituent in the foundry of power that forged the revolutionary outcomes. In spite of themselves, the Bolsheviks accepted bureaucracy as one of their tools for welding together a postrevolutionary social and political framework. I mean this not only in the sense that I take Theodore H. Von Laue to be arguing: that is, that state and Party bureaucracies "administered" or imposed revolutionary policies and continued state initiatives in socioeconomic development.[21] I mean that bureaucracy became a tool for change in the sense that it served as a means of transposing class roles. In Soviet civil administration after 1917, the meek indeed inherited the earth.

---

20. This view of the early development of post-revolutionary administration is often advanced by Soviet historians: e.g., E. G. Gimpel'son, *Velikii oktiabr' i stanovlenie sovetskoi sistemy upravleniia narodnym khoziaistvom (noiabr' 1917–1920gg.)* (Moscow, 1977), and *Rabochii klass v upravlenii sovetskim gosudarstvom: Noiabr' 1917–1920 gg.* (Moscow, 1982); A. A. Melidov, *Istoriia gosudarstvennykh uchrezhdenii SSSR, 1917–1936 gg.: Uchebnoe posobie* (Moscow, 1962). The same view is not uncommon among Western scholars: see, e.g., Theda Skocpol, *States and Social Revolutions: A Comparative Analysis of France, Russia, and China* (New York, 1979), 226ff.; Bendix, *Nation-Building*, 185; Samuel P. Huntington, *Political Order in Changing Societies* (New Haven, Conn., 1968), esp. the treatment of Leninism and Bolshevism in the Russian Revolution of 1917, 334ff.

21. Theodore H. Von Laue, *Sergei Witte and the Industrialization of Russia* (New York, 1969), chap. 3; *Why Lenin? Why Stalin? A Reappraisal of the Russian Revolution, 1900–1930* (Philadelphia, 1964), chaps. 9, 10; and "Imperial Russia at the Turn of the Century: The Cultural Slope and the Revolution from Without," *Comparative Studies in Society and History* 3 (1960–61), 353–367.

Bureaucracy and Revolution

Clearly, there are important ways in which bureaucrats and bureaucracies in Russia contributed not only to economic development but to revolution. For one thing, certain parts of the bureaucratic apparatus are sometimes—frequently, in the case of Russia—at the technological cutting edge in developing societies. As employers, bureaucracies monopolize a large share of the technically educated elite. And in societies such as Russia, where bureaucracy in the nineteenth century tended to expand more rapidly than private enterprise, bureaucratic technical groups acquired an importance that was even greater than their very substantial numbers would suggest.[22]

Second, while it is true that specialists often remain in subordinate bureaucratic roles everywhere—and especially in Old Regime Russia—their class background makes them readier participants in radical change than their official superiors. The most important source of officials who were holdovers from the tsarist years was the socially inferior sub-elite that possessed special skills of use to the new government. For these segments of Russian society, revolution created opportunities for advancement that had been unthinkable under the Old Regime.

Third, the greater tendency of non-noble bureaucrats to adopt specialist roles ensures their further professionalization as career bureaucrats; that is, they more readily associate themselves with social groups that are cut off from traditional status as expressed by landownership or social eliteness. It is precisely such professionalized bureaucrats whom Ellen Kay Trimberger and Theda Skocpol identify as critical to "revolution from above" and who give the central political apparatus the administrative resources necessary for cohesion, independence of action, and, ultimately, survival during the anarchy of revolution.[23] In the case of Russia, let it be emphasized, the focus on the *revolutionary* roles of these sub-elites should not distract attention from the fact that they served as a foundation for central political power over long periods of time—not merely in episodes of revolution, whether from above or

---

22. Olga Crisp, *Studies in the Russian Economy before 1914* (London, 1976), 22–34; Alexander Gershenkron, *Continuity in History and Other Essays* (Cambridge, Mass., 1968), 140–248; A. J. Rieber, *Merchants and Entrepreneurs in Imperial Russia* (Chapel Hill, N.C., 1982), 3–132.

23. Ellen Kay Trimberger, *Revolution from Above: Military Bureaucrats and Development in Japan, Turkey, Egypt, and Peru* (New Brunswick, N.J., 1979), and "A Theory of Elite Revolutions," *Studies in Comparative International Development* 7 (1972), 191–207; Theda Skocpol, *States and Social Revolutions*, and "France, Russia, China: A Structural Analysis of Social Revolutions," *Comparative Studies in Society and History* 18 (1976), 175–210.

below. Under certain circumstances, I argue, segments of the civil bureaucracy will function as supports and resources for revolution but, paradoxically, serve as a mainstay of central political power against the centrifugal forces of rural and commercial self-interest, thereby accounting for some of the principal features of the revolutionary outcome.

Fourth, precisely because of the conjunction of lower-class status, bureaucratic employment, and specialist education or experience, bureaucratic sub-elites form part of the prow, the cutting edge, of social revolution, drawing in their wake by their successful example other segments of society previously excluded from clerical status. Following the revolution in Russia, at least, employment in the bureaucracy was one of the quickest routes for thousands to the fulfillment of the revolutionary promise of social uplift.

Fifth and finally, this study shows a continuation and even an intensification of the relation between social and bureaucratic authority status that existed in the Old Regime bureaucracy.[24] Because of the enduring connection between the privileges of birth and the power of office, there was an impersonal, structural sense in which the bureaucracy served—unconsciously, of course—as an implement of revolution in the provinces. The role of the landowning nobility in provincial and local bureaucracy was so important that when the noble landowners were dispossessed after 1917, the destruction of the apparatus of provincial and local administration occurred simultaneously. Here, even more than at the national level, one finds social revolution closely bound up with bureaucratic revolution.

Agency and Structure

Questions of control, responsibility, causation, and human agency in large-scale political and social events bulk exceptionally large in the history of Russia and the USSR. In part, this is because individuals have played enormously important roles throughout Russian history even up to the present time. Personalities—Ivan Groznyi, Peter the Great, Alexander II, Stalin—so dominate the language we use to explain events in Russian history that it is mandatory to comment, if only briefly, on the interpretive difficulties such language raises.

24. This view appears to contrast with that of Roberta Thompson Manning, *The Crisis of the Old Order in Russia: Gentry and Government* (Princeton, N.J., 1982), 2–64. Manning sees the emergence of the nobles' interest in their personal estates and local affairs as a rejection of the professionalized bureaucracy. For an extended critique of Manning's work, see Seymour Becker, *Nobility and Privilege in Late Imperial Russia* (DeKalb, Ill., 1985). Also see my "Structure, Class, and Career: The Problem of Bureaucracy and Society in Russia, 1801–1917," *Social Science History* 6, no. 1 (1982), 87–110.

Let us look at an illustration. One constraint that weighed heavily on Bolshevik chances for success was the aggregate of human resources to which any government could lay claim. These resources were essentially given before the outbreak of revolution. Thus, for example, Trotsky necessarily approached the organization of the Red Army as an effort to work within the very short time frames imposed by war and diplomacy, using highly limited social and demographic resources.[25] Similarly, the human resources available to the civil administration—the number, quality, and experience of personnel together with their collective attitudes toward the new state apparatus—were essentially given long before the Bolsheviks came to power. The choices made by persons in authority were always shaped by factors over which they had little or no control and which they often understood but poorly. Under these circumstances it is hardly likely that Trotsky or anyone else could have made rational and competent decisions about the formation and use of the Red Army. The behavior of individuals, in short, frequently either misleads us or provides only fragmentary explanations for historical events.

This book relies on several alternative sources of explanation. As a rule, the political and administrative structures either of the Empire or of the Soviet era are described and evaluated first in the effort to arrive at an understanding of the framework and the impetus for social transformation. The roles of individuals generally occupy a secondary level of explanation. Since change in large, formal organizations is my principal focus, it is often more convenient and frequently more accurate to use the term "structure" in formulating descriptions and analyses than to use the behavior of political actors or class motivations.

"Structure," of course, has long been a useful term in both historical and sociological analysis. In the language of some earlier sociologists, organizational structure seemed almost something one could touch, and the broad social importance of fitting into the preexisting structure could hardly be understated: "No individual becomes a moral person", said Everett C. Hughes, "until he has a sense of his own station and the ways proper to it."[26]

But "structure" has its own history as an analytic tool and as a concept, and it is afflicted with its own particular definitional problems.

---

25. Isaac Deutscher, *The Prophet Armed: Trotsky, 1879–1921* (New York, 1965), 405–415; Marc Ferro, "The Russian Soldier in 1917: Undisciplined, Patriotic, and Revolutionary," *Slavic Review* 30 (1971), 483–512; Allan K. Wildman, *The End of the Russian Imperial Army: The Old Army and the Soldiers' Revolt (March–April, 1917)* (Princeton, N.J., 1980); L. G. Protasov, "Klassovyi sostav soldat russkoi armii pered oktiabrem," *Istoriia SSSR* 1 (1977), 33–48.

26. Everett C. Hughes, "Institutional Office and the Person," *American Journal of Sociology* 43 (1937), 404.

Since the beginning of the twentieth century, structural analysis has served as possibly the single most important methodological technique in comparative history.[27] At the same time, it has been employed by historians who are sensitive to the need to avoid the distortions of personality-dominated historical interpretations.[28] As argued by Claude Lévi-Strauss a generation ago, however, under detailed analysis the inflexible dimensions of the dichotomy of structure and function become less and less substantial; what previously appeared to be fixed or complete "structures" became, under close observation aggregates or patterns of action, constantly evolving.[29] When disassembled, or "deconstructed," into their component behaviors, previously acknowledged "structures" dissolve, together with their structured "history," or "conjunctures"; and both cross-time and cross-cultural dynamic parallels become more evident and instructive. While structuralist terminology was familiar in the phenomenology, linguistic analysis, and history of the pre–World War II era, it has now been superseded by a rhetoric of "deconstruction" in literary and linguistic analysis and of "conjunctures" in historical studies. This rhetoric, nevertheless, still owes much to its structuralist origins.[30]

---

27. Wolfgang J. Mommsen, "Die Geschichtswissenschaft in der modernen Industriegesellschaft," *Vierteljahrshefte für Zeitgeschichte* 22, no. 1 (1974), 1–17; Alette O. Hill and Boyd H. Hill, Jr., "Marc Bloch and Comparative History," *American Historical Review* 85, no. 4 (1980), esp. 843–45; Theda Skocpol and Margaret Somers, "The Uses of Comparative History in Macrosocial Inquiry," *Comparative Studies in Society and History* 22, no. 2 (1980), 174–97.

28. Skocpol, "France, Russia, China," 175–210; D. Groh, "Strukturgeschichte also 'Totale' Geschichte," *Vierteljahrschrift für Sozial- und Wirtschaftsgeschichte* 58, no. 3 (1971), 289–322; Peter Manicas, "Review Essay: *States and Social Revolutions*," *History and Theory* 22, no. 2 (1981), 204–18.

29. Claude Lévi-Strauss, *Structural Anthropology* (New York, 1963), 31–35. See also Lévi-Strauss, *The Raw and the Cooked*, trans. John and Doreen Weightman (New York, 1970), introduction ("Overture"), and the remarkable commentary on it by Jacques Derrida, "Structure, Sign, and Play in the Discourse of the Human Sciences," in Richard Macksey and Euginio Donato, eds., *The Structuralist Controversy: The Languages of Criticism and the Sciences of Man* (Baltimore, Md., 1970), 247–65.

30. Mark Philp, "Foucault on Power: A Problem in Radical Translation?" *Political Theory* 11, no. 1 (1983), 29–52; Peter Manicas, "The Concept of Social Structure," *Journal of the Theory of Social Behaviour* 10, no. 2 (1980), 65–82; Philip Abrams, "History, Sociology, Historical Sociology," *Past and Present* 87 (1980), 3–16, in which the references are to "eventuation" and "structuring." See Georg G. Iggers, "Die 'Annales' und ihre Kritiker: Probleme moderner franzosischer Sozialgeschichte," *Historische Zeitschrift* 219, no. 3 (1974), 578–608. On "conjunctures" in history, see Fernand Braudel, *The Perspective of the World*, vol. 3 of *Civilization and Capitalism, 15th–18th Century* (New York, 1984), 71–85. The links between comparable concepts of structure and Marxist thought are explored in John Ryan, *Marxism and Deconstruction: A Critical Articulation* (Baltimore, Md., 1980), esp. chaps. 1, 8, 9; Göran Therborn, *Science, Class, and Society: On the Formation of Sociology and Historical Materialism* (London, 1980), chap. 6; and E. M. Shtaerman, "On the Problem of Structural Analysis of History," *Soviet Studies in History* 7, no. 4 (1969), 38–55 (a translation of his article in *Voprosy istorii*, no. 6 [1968]).

In this book, when an explanation is couched in terms of "structure," the reference is to established patterns of behavior; to oversimplify, the reference is to the modal values of that behavior. Law, oral tradition, patterns of career advancement, even the physical arrangements of offices and the symbols of different degrees of organizational status are treated as the evidence of structure rather than structure itself. Their connection with organizational behavior, in my view, is that of historical evidence.

When I refer to the importance of structure as determining those factors with which the Bolsheviks had to contend after October 1917, I have in mind a large body of patterns of behavior within formal bureaucratic organizations and between those organizations and the rest of society. These patterns are represented or measured by social variables such as class background, education, and family characteristics, but they also include organizational variables such as professional experience, specialization, career patterns, and organizational status. The components of structure also include groups of variables that define the relations between the organization and society, such as formal authority, power, and the degree to which the organization monopolizes resources that are beyond organizational control in other societies. In this context, of course, the tendency toward technocracy is a product of some structures and a part of others. It is a product both of political and administrative will and also of preestablished administrative and political structures with which the current generation has to contend.

In this light, then, the problem of this study, in Abrams's words, is "the problem of finding a way of accounting for human experience which recognizes simultaneously and in equal measure that history and society are made by constant, more or less purposeful, individual action and that individual action, however purposeful, is made by history and society."[31]

In his famous antiestablishment short story "How a Muzhik Fed Two Officials," Mikhail Saltykov expresses the same complex relationship simply and elegantly. Having endured the extraordinary—one could say "revolutionary"—experience of magical transportation to a deserted isle, two *chinovniki* (Imperial officials) find themselves without food or any of the other comforts and privileges to which their successful careers had accustomed them. They are saved when they happen on the solution of locating a *muzhik* (peasant) to serve them. The officials

---

31. Abrams, "History, Sociology, Historical Sociology," 7. Although it is chronologically organized in a broad sense, this book is not a narrative history; I do not aim here to tell a story.

will not change their pretentious and extravagant tastes merely because of changed circumstances, and they certainly have no intention of engaging in manual labor themselves. So they force the *muzhik* to serve their needs. It gradually becomes clear, however, that while the officials parasitize the peasant, he accepts their power and status with no reason to do so except the force of tradition: "He had to submit to his fate. He had to work." At the end of his hard day of service, the *muzhik* is tired and asks his masters, "Will you permit me to rest a little?" The *chinovniki* reply:

> "Go take a little rest, but first make a good strong cord."
> The muzhik gathered wild hemp stalks, laid them in water, beat them and broke them, and toward evening a good stout cord was ready. The officials took the cord and bound the muzhik to a tree, so that he could not run away. Then they laid themselves to sleep.[32]

As time passed in Russia, social, economic and political conditions changed—slowly and relentlessly for a long time and then, in the revolution, suddenly and dramatically. Eventually the class actors and their respective roles were to change too. *Muzhiks*, indeed, would enter the ranks of officials. Although we shall look closely at this "revolutionary" exchange of roles, our data make this part of our study relatively simple. Far more interesting, complex, and important will be our effort to understand how the structure of relations between society and administration by turns endured and changed.

32. Mikhail Saltykov, "How a Muzhik Fed Two Officials," in Norris Houghton, ed., *Great Russian Short Stories* (New York, 1958), 91.

CHAPTER 2

Domestic Administration in Structural Perspective, before 1917

This book aims to construct its arguments and conclusions on the basis of information about administrative structure and personnel. Detailed descriptions of the organizational development of Russian administration are readily available to the reader.[1] Thus, the objective here is merely to summarize specific major trends that are particularly interesting for the purposes of this study. These trends include structural change in ministerial government in the nineteenth century, the changing relation between social class and the organizational status of officials, and status and career differences among officials. I construct this summary by focusing on specific elements of ministerial structure, mainly the offices and personnel, of nineteenth-century civil administration. The necessary information groups itself in three major eras between 1800 and 1930. And although my focus is on the last forty years of this period, the need for historical perspective requires a longer preliminary view.

1. N. P. Eroshkin, *Istoriia gosudarstvennykh uchrezhdenii dorevoliutsionnoi Rossii* (Moscow, 1983), 138–251; Hans-Joachim Torke, "Das russische Beamtentum in der ersten Hälfte des 19. Jahrhunderts," *Forschungen zur osteuropaeischen Geschichte* 13 (1967), 7–345; Erik Amburger, *Geschichte der Behördenorganisation Russlands vom Peter dem Grossen bis 1917* (Leiden, 1966); George Yaney, *The Systematization of Russian Government* (Urbana, Ill., 1973), 193–380; P. A. Zaionchkovskii, *Pravitel'stvennyi apparat samoderzhavnoi Rossii v XIX v* (Moscow, 1978).

The Creation of Ministerial Government

Russian ministerial government began in 1801, and from then until the early 1830s it passed through several stages of modification in detail which finally produced a standard ministerial structure. This fact is important because it is during this period that the outlines of modern Russian and Soviet administration can be discerned. All of the characteristics not only of Imperial but of Soviet ministerial administration that are familiar today are found, at least in outline, between 1801 and 1835.

The features of contemporary administration present during the early 1800s include the central ministries themselves, each with its single chief—the minister—and his deputies; an advisory council (a kind of minister's cabinet) composed of senior officials within the agency; a secretariat or chancellery; and a variable number of specialized departments that could be added to or subtracted from the ministerial apparatus, depending upon changes in its legal grant of authority. This particular combination of offices was not, of course, unique to Russia; the important point is that it produced an organizational flexibility that was attractive to political authorities. The Ministry of Internal Affairs, for example, went through many major changes in personnel and office complement (*sostav*) in the nineteenth century, but the ministry itself endured throughout.

This same approach to organization concentrated specialized personnel on particular problems in a way that has served Russian administration reasonably well ever since. The problem of who would control these specialists was solved variously in the nineteenth and twentieth centuries because the ministerial structure was sufficiently flexible to admit all the different approaches to control and still ensure that specialists were available. Even in the wild days of the 1920s, department heads would sometimes carry their teams of specialists with them from one organizational setting to another. The outlines that emerged in the early 1800s also included a parallel structure, increasingly elaborate over time, of ministerial subdivisions at the provincial and subprovincial (or, in a later era, republic, *oblast*, and sub-*oblast*) levels.[2]

Legislation between 1801 and 1811 created several "ministers" and

---

2. *Svod Zakonov Rossiiskoi Imperii*, Tom III, Izdanie 1896 g. *Uchrezhdenie ministerstv* (St. Petersburg, 1896, 1912, 1913, 1914 gg.); T. H. Rigby, *Lenin's Government: SOVNARKOM, 1917–1922* (Cambridge, 1979), chaps. 9–11; A. P. Kositsyn et al., eds., *Istoriia Sovetskogo gosudarstva i prava: Stanovlenie Sovetskogo gosudarstva i prava (1917–1920 gg.)* (Moscow, 1968), chaps. 3, 4; A. A. Melidov, *Istoriia gosudarstvennykh uchrezhdenii SSSR, 1917–1936 gg.: Uchebnoe posobie* (Moscow, 1962).

provided them with administrative offices.³ Although N. P. Eroshkin refers to the creation of "ministries," in fact ministries were not initially established as independent organizations.⁴ Instead, the legislation produced a complicated amalgam of the eighteenth-century administrative system of Peter the Great and the more streamlined, centralized "offices of ministers" that ministries were destined to become.⁵ The principal offices included War, Marine, External Affairs, Internal Affairs, Commerce, Finances, Public Education, and Justice. But as the conservative historian Karamzin noted, this attempt was rather premature, made "with excessive haste." The new offices were "created and set in motion before the ministers had been provided with an Instruction" or a clear "guide to help them carry out their important duties!"⁶ Subsequent legislation shifted functions around, depending on the changing objectives of the chief political authorities of the state and even upon their moods, and the names of some organizations changed frequently in the nineteenth century. But the basic structure emerged rapidly and has endured through a remarkably long and varied time.

The power and responsibility of the minister himself represented a major departure from previously established forms of administration. Part of this power seems only symbolic, but it was much more than symbolism. The minister enjoyed the legal right to address the emperor directly and individually in his annual *vsepodanneishie doklady* ("most devoted") official reports. By the intensity of his scorn Karamzin recognized the threat that this access posed to already established institutions. The ministers, he growled, "having emerged upon the ruins of the colleges... wedged themselves between the sovereign and the people, eclipsing the Senate, and divesting it of its power and greatness. And, although the ministers are subordinate to the Senate insofar as they must submit reports to it, yet by being able to say, 'I had the pleasure to report to His Majesty!' they can silence the Senators, with the result that so far this alleged responsibility has proven but a meaningless ritual."⁷

Karamzin demonstrated both his political sensitivity and his prescience. The social prestige and political influence acquired through access to the imperial person did result in rapid enhancement of the ministers' authority, singling them out of the mix of officers of more

3. PSZ, 1802, no. 20406; 1811, no. 24686. Other legislative references may be found in *Svod Zakonov, Tom III, Uchrezhdenie ministerstv*
4. Eroshkin, *Istoriia gosudarstvennykh uchrezhdenii*, 156.
5. Yaney, *Systematization*, 193–212.
6. Richard Pipes, ed., *Karamzin's Memoir on Ancient and Modern Russia* (Cambridge, Mass., 1966), 149.
7. Ibid., 150; but see Yaney's description (*Systematization*, 301–5) of the irregularity of report-making.

ancient lineage and those, such as the senators, whose roles had been the product of Peter I's administrative reforms. At the same time this authority, combined with their organizational flexibility and a capacity to focus on many specific functions, made the ministers powerful as a group and a threat to the autocracy itself. The recognition of this threat, which would have its echo in the competition between political generalists and technocratic specialists of a later era, was already evident in the reluctance of the autocracy to permit the creation of a formal ministerial cabinet and a prime minister.

Just one hundred years before, Peter the Great had attempted to cram his upper bureaucracy, the *kollegii*, into a single building where he could keep an eye on them and limit their pretensions; but soon after the publication in 1802 of Alexander's laws, imposing neoclassical office blocks destined to house the new ministries began to make their appearance, one by one, within sight of the Winter Palace.[8]

It is true that particular persons in state administration, such as the emperor's friend Victor Kochubei, a minister of internal affairs, continued to be important. Nevertheless, stress must be laid on the formation of elaborate, semiindependent bureaucracies, agencies that in principle were organized not around particular persons but around functions that could be subdivided almost infinitely into operationally specific departments. Paradoxically, this structure did not mean that ministerial government produced a system of administrative isolates. The opposite was true: although their political ambitions were often divergent, the ministers shared a broad set of purely administrative concerns that were based on the similarity both of ministerial organization and, increasingly, of concerns over personnel. The frequent gatherings of the ministers, whether in the Committee of Ministers or in other forums, was a source of power for the new system, and this fact was recognized by both political insiders and those who watched them.[9]

The creation of a new coordinated administrative system eventually required the development of an elaborate set of uniform rules governing the conditions both of official employment and official behavior within the various organizations—the *Ustav o sluzhbe*, or statute on service. To be sure, some such rules had existed before the early nineteenth century, but the legislation of the 1820s and 1830s was an attempt to create service organizations that would deal in parallel manner with

---

8. *Ocherki istorii Leningrada* (Moscow, 1955), 1:550–600, 2:793–809.
9. Yaney, *Systematization*, 194. Also see Richard Wortman, *The Development of a Russian Legal Consciousness* (Chicago, 1976), chap. 2, "Bureaucratization, Specialization, and Education."

each of the major tasks of government—law enforcement, tax collection, defense, regulation of commerce, and so forth.[10]

Taken as a whole, the body of ministerial legislation gave evidence of both a logically and flexibly subdivided administration and of a serving body whose organizational roles and conditions of work would be uniformly coordinated. This coordination, moreover, emphasizes the fact that *service* itself was becoming important.

Legislation governing promotion had long provided for some degree of equivalence between military and civil ranks, thereby making transfer, or seconding, possible. In the nineteenth century the military and civilian services became more distinct from each other, but the parallel ministerial structures within the civil service made uniform the relation between rank and office throughout the major administrative arms of the civil bureaucracy and introduced what would become a universal preoccupation with career, *organizational* status, and seniority.[11] The legislation on ministries and the parallel statutes on personnel sounded the first somber chords of a requiem for the era when civil service was merely a stepping stone to social and political preferment or an obligation imposed upon *une noblesse obligée*. During the lifetimes of the middle generations of the nineteenth century, career would become more and more important both for ambitious lower classes and for the nobility, even the landed nobility.

A lot has been made of the vicissitudes, the outright unpredictability and arbitrariness, of Russian administration and of the administrative career.[12] Hundreds or even thousands of cases support the idea that Russian bureaucracy was arbitrary, corrupt, and prone to political intervention; at the same time, tens of thousands of cases, endlessly summarized in career dossiers (the *formuliarnye spiski*), indicate the opposite. System, eternally repeated tasks, impersonality, and organizational demands were all combining to form the rigid walls of the

---

10. PSZ, 1 August 1801, no. 19961; 4 February 1803, no. 20608; 2 June 1808, no. 23056; and especially 25 June 1834, no. 7224 (statutes). For the entire code, see *Svod Zakonov*, Tom III, *Ustav o sluzhbe po opredeleniiu ot pravitel'stva i polozhenie ob osobykh preimushchestvakh grazhdanskoi sluzhby v otdalennykh mestnostiakh*.

11. *Svod Zakonov*, Tom III, *Ustav o sluzhbe po opredeleniiu ot pravitel'stva i polozhenie ob osobykh preimushchestvakh grazhdanskoi sluzhby v otdalennykh mestnostiakh* (1915), divisions 1 and 2, 1–215; also PSZ, 1811, no. 24686, "Obshchie polozheniia ministerstv," which specifies the size and kind of administrative staff to which every ministry would have access.

12. See, e.g., Marc Raeff, "The Russian Autocracy and Its Officials," *Harvard Slavic Studies* 4 (1957), 77–91. For a contemporary judgment, see V. D. Kuz'min-Karavaev, *Zemstvo i derevnia, 1898–1908: Stat'i, referaty, doklady i rechi* (N.p., n.d.); V. V. Veselovskii, *Istoriia Zemstva za Sorok Let* (St. Petersburg, 1911); Leonid Dashkevich, *Nashe Ministerstvo Vnutrennikh Del* (Berlin, 1895).

mold of career. The humdrum predictability of bureaucratic life became the bane of the intelligentsia and nobility alike in nineteenth-century Russia.

With the creation of ministries, state administration as a whole began a sustained expansion not only of offices but of personnel. That expansion would continue right through the nineteenth century.[13] Significantly, the growth of both organization and personnel appears to have outstripped the ability or the willingness of the nobility to provide candidates for official positions and, fairly rapidly, education became a better predictor of career advancement than upper-class status as measured by landholding or social origin alone.[14] Thus, we can date in this period the beginning of the erosion of the Russian landed aristocracy's position as the unquestioned elite of the administrative apparatus, and the rising competition for that title from an educated, professional administrative elite—members of the upper class, to be sure, but not necessarily of the nobility.

This emergence in the nineteenth century of successful bureaucrats who were not members of the noble estate ran parallel to the advance of specialists or technocratic administrators. To some extent, moreover, these two developments were responses to the same forces of modernization, though neither of them would reach either its logical or historical climax before 1917. In particular, the landed elites in Russian government would, from time to time, find ways to manipulate the system in order to hang on to positions of exceptional power until the very end. This important fact is another reason for giving full recognition to the significance of system and predictability within the profession of Russian state administration in the nineteenth century. Otherwise, we miss the point that it was possible for the noble landed elite to accept a system that was in many ways repugnant to them and still, sometimes, to manage it to their collective benefit.

The Introduction of Technical Administration

Starting about midcentury, Russian administration underwent a further reorientation that was just as important as the creation of minis-

---

13. S. F. Starr, *Decentralization and Self-Government in Russia, 1830–70* (Princeton, N.J., 1972), 3–50.
14. Walter M. Pintner, "Evolution of Civil Officialdom," 190–226, and "Civil Officialdom and the Nobility in the 1850s," 228–49, both in Walter M. Pintner and Don Karl Rowney, eds., *Russian Officialdom: The Bureaucratization of Russian Society from the Seventeenth to the Twentieth Century* (Chapel Hill, N.C., 1980); Jerome Blum, *The End of the Old Order in Rural Europe* (Princeton, N.J., 1978), 421; Seymour Becker, *Nobility and Privilege in Late Imperial Russia* (DeKalb, Ill., 1985), 108–12.

terial government early in the century. Officials with special technical training—physicians, statisticians, veterinarians, engineers—began to make their appearance in force.

The arrival of the first specialists was combined with a rapid extension of offices and personnel generally. With the emancipation of Russia's serfs mainly in the 1860s, it was necessary to replace the landlords as sources of local authority for an enormous segment of the Russian population: the inhabitants of villages and towns. In addition, even if we do not interpret these developments as a "compulsion" (to use George Yaney's term) to attack the peasantry, it was necessary to correct decades, if not centuries, of the relative neglect of provincial administration typical of the urban, cosmopolitan preoccupation of great land empires.[15] The result was an expansion of the flexible ministerial system into regional and especially local government. That the system survived and even flourished under this expansion is a tribute to its inherent adaptability, for the changes consisted in far more than a simple elaboration of the familiar activities of the central government.[16]

From midcentury, Russian local and provincial government included for the first time administrative positions that had not previously existed or were to be found only in the central government. Moreover, through the creation of rural land assemblies (zemstvos), these positions involved individuals—including impoverished and poorly educated noble smallholders and even peasants—who had never previously been included in official roles in Russian government. For an understanding of the importance of these changes, it is helpful to trace briefly the development of the largest and most important ministry.

The Ministry of Internal Affairs first saw the light of day under instructions to maintain law and order and to secure the peace and welfare of the population. This alignment of responsibilities, in its comprehensive range, made the ministry a kind of "welfare bureaucracy" charged with the oversight, care, and development of broad, vague

---

15. On the "compulsion," see Yaney, *Systematization*, 239–42. On administrative neglect, see Starr, *Decentralization*, 3–109. The urban cosmopolitan focus of great land empires is described in S. M. Eisenstadt, *The Political Systems of Empires: The Rise and Fall of Historical Bureaucratic Societies* (New York, 1963). See also Eisenstadt's *Revolution and the Transformation of Societies: A Comparative Study of Civilizations* (New York, 1978).

16. Starr, *Decentralization*, 110–74; Terence Emmons, *The Russian Landed Gentry and the Peasant Emancipation of 1861* (Cambridge, 1968), 209–65; Eroshkin, *Istoriia gosudarstvennykh uchrezhdenii*, 193–251; P. A. Zaionchkovskii, *Otmena krepostnogo prava v Rossii* (Moscow, 1960). For legislation affecting the Ministry of Internal Affairs, see *PSZ*, 1861, no. 3729; 1863, no. 39858. For the Ministry of Education, see *PSZ*, 1863, no. 39751.

segments of the public as well as of its own officials.[17] So broad an assignment prompted Karamzin to observe that it could not be discharged: "Was the Minister of the Interior, who appropriated for himself nearly all of Russia, capable of gaining a good insight into the endless stream of papers flowing through his office? Could he understand at all subjects of such diversity?"[18]

Keeping the peace, mainly in urban areas, was the ministry's most typical assignment. But this task was combined with maintaining postal communications, monitoring public health, licensing physicians, and many other functions, many of which required the presence of some personnel with special skills. In 1836, medical administration became a major Internal Affairs function; in 1863 the Central Statistical Administration was added; in 1868, the Veterinary Administration.[19] After emancipation of the state's and landlords' serfs in the 1860s, moreover, the ministry became principally responsible for arranging the details of land division (between landlords and peasants) and for overseeing rural order and welfare thereafter.[20] Of course, the extension of administrative authority to the countryside in the 1860s also involved extending the welfare or well-being (*blagosostoianie*) obligations, which increased not only in number but—in order to deal with the increasingly technical functions of a rapidly growing and increasingly complex society—in technical sophistication.

The influx of new technical roles, however, was not only the result of expanding regional and local administrative responsibilities. There were also important related changes in the world beyond both the village and the boundaries of official Russia. The industrial revolution in England and western Europe intensified the demand for raw materials, including food, from less developed countries.[21] As a consequence, Russian exports rose in value from 181.3 million rubles (Rs) in 1860 to 716.2 million by 1900. Typically, a large portion of this export trade was accounted for by agriculture.[22]

---

17. Robert V. Presthus, "Weberian vs. Welfare Bureaucracy in Traditional Society," *Administrative Science Quarterly* 6, no. 1 (June 1961), 1–24.
18. Pipes, *Karamzin's Memoir*, 150.
19. PSZ, 1836, no. 9317; 1863, no. 39566; 1868, no. 46500; 1870, no. 48576.
20. Yaney, *Systematization*, 187–192, 306–7; Zaionchkovskii, *Pravitel'stvennyi apparat*, 179–220.
21. Simon Kuznets, *Economic Growth and Structure* (New York, 1965), 250–56; Paul Bairoch, "Niveaux de developpement économique de 1810–1910," *Annales: Economies, Sociétés, Civilisations*, 1965, pp. 1091–1117; R. W. Goldsmith, "The Economic Growth of Tsarist Russia, 1860–1913," *Economic Development and Cultural Change* 9, no. 3 (1961), 441–75.
22. Olga Crisp, *Studies in the Russian Economy before 1914* (London, 1976), 112, 97–108; Goldsmith, "Economic Growth," 450 n.19; Central Statistical Administration, *Statisticheskii ezhegodnik Rossii* (St. Petersburg, 1893–1914).

Domestic Administration before 1917 27

In addition to the importance of agriculture in foreign trade, of course, was the obvious fact that an insufficient food supply could and did frequently cause distress in the expanding population of rural Russia.[23] Therefore, Russian officialdom was concerned, preoccupied, with such questions as the state of the harvest, the market price of the various grains, the cost of agricultural labor and land.

To respond to these concerns, the Ministry of Internal Affairs was obliged to develop an extensive capacity to gather and to analyze statistics.[24] These activities culminated, in a sense, in the first all-Russian census in 1897, by which time the Central Statistical Administration employed one of the ministry's—and, no doubt, the entire country's—largest concentrations of technical specialists.

To be sure, the increase in technical roles was not limited to Internal Affairs; indeed, it was in even greater evidence elsewhere. For example, in 1811 roads and other internal communications became the responsibility of the Main (or Chief) Administration of Ways of Communication, an institution with approximately the same degree of authority as a ministry but lacking the ministry's legal status. The development of railways necessitated its restructuring in 1865 into a technically more sophisticated Ministry of Ways of Communication, which assumed responsibility for training its own engineers and included one of the highest concentrations of specialists in the government.[25] It is also true that the introduction of technically trained officials into Russian service began much earlier than midcentury.[26] Still, it is from the 1860s that we can see a trend involving the entire government, including competing agencies such as the ministries of Finances and Internal Affairs.

To continue with the example of Internal Affairs, its structure as it had evolved by the early twentieth century, a hundred years after its creation, reveals a very substantial transformation. The principal characteristic of this transformation, moreover, was expansion, a characteristic it shared with other ministries. By 1900 it consisted of some twenty-five separate agencies, and the number of ranking civil servants in the central ministry totaled about a thousand.[27] Its annual operating

---

23. Richard G. Robbins, Jr., *Famine in Russia, 1891–1892: The Imperial Government Responds to a Crisis* (New York, 1975), 1–13. But see also James Y. Simms, Jr., "The Crop Failure of 1891: Soil Exhaustion, Technological Backwardness, and Russia's 'Agrarian Crisis,' " *Slavic Review* 41, no. 2 (1982), 236–50.
24. PSZ, 1863, no. 39566; 1875, no. 54742.
25. Eroshkin, *Istoriia gosudarstvennykh uchrezhdenii*, 216–17.
26. John A. Armstrong, *The European Administrative Elite* (Princton, N.J., 1973), 178–79.
27. Ruling Senate, *Adres-Kalendar': Obshchaia rospis nachal'stvuiushchikh i prochikh dolzhnostnykh lits po vsem upravleniiam v Rossiiskoi Imperii na 1904 god*, pt. 1 (St. Petersburg, 1904), 133–62.

budget (including salaries) exceeded Rs 2.2 million. And the single most impressive group of additions during the nineteenth century was certainly concentrated in the professional-technical areas already mentioned.[28] Indeed, the Ministry of Internal Affairs, like other Russian ministries, and still following the welfare bureaucracy pattern, directly developed expertise, establishing its own in-house training and research institutes in medicine, civil engineering, communications technology (or, as it was then called, electrotechnical science), and statistics.[29]

The Early Transition to Technocracy

The third era of change with which we are concerned began with the Revolution of 1917 and extended through the end of the 1920s. This period saw the completion of the structural changes that produced the foundations of Soviet technocracy. Although ministers were now called "commissars"—a throwback to the administrative terminology of the early eighteenth century—and ministries duly became "commissariats," major structural characteristics endured. Commissars still ran their organizations with the assistance of deputies and were advised by councils (now *kollegii* instead of *sovety*); their operations were distributed among functionally coherent and frequently highly specialized departments.

Meanwhile, the attempt to eradicate private property and to force the state to assume responsibility for wide new areas of public welfare— such as all forms of medical care—resulted in a rapid expansion of administrative roles similar to that resulting from the eradication of serfdom and landlord power in the 1860s. Administration eventually extended to functions and areas in which it had never before been involved, such as comprehensive medicine, social welfare, and very broad-gauged yet detailed economic administration—operational areas where specialists were indispensable.

Another highly significant change was the first major rupture in at least a century of the rules and customs governing induction and promotion, a departure from established precedent very important in its larger implications. The upper-class elites who had dominated civil administration since its inception were finally swept away. They were replaced by former civil servants of inferior organizational and social status and thousands of workers and peasants in the most dramatic

28. *Smeta dokhodov i raskhodov i spetsial'nykh sredstv Ministerstva Vnutrennikh Del na 1905 g.*, data "naznachenie 1904 g." (St. Petersburg, 1904).
29. Armstrong, *European Administrative Elite*, 223ff.

social transformation of government administration in modern European history. This rupture of "normal" intergenerational change deprived the nobility of its influence, such as it was, in state administration. Further, it brought to an end a long tradition that had provided employment for the nobles and assured them of preferment once employed.

The break in established patterns of personnel change was important as well for less obvious reasons. The principal components of an organization are its personnel; they form a quasi-society in some ways much like society in the "outside" world, in some ways different. In particular, there is often a close relationship between seniority (measured by chronological age or length of service) and authority or perceived power in the organization. Older people—those with more organizational experience, less up-to-date education, a tendency to conservatism—often make decisions that are organizationally important for their younger and less experienced colleagues: whether to hire or fire, whether to promote, how long to retain, and generally how to reward or punish an individual as well as what behavior to reward or to punish. When such a pattern is broken—for example, when senior staff are dismissed or resign en masse (as happened in Russia in 1917 and 1918), or when junior staff promotions are delayed beyond expectations (as in the 1960s and 1970s in the USSR)—the disruption will presumably always be substantial. In the one case, experience and authority are sacrificed; in the other, the energy and ambition of youth are lost. This latter circumstance probably explains some of the sharp contrasts between the Soviet administrative innovativeness, even recklessness, of the 1930s and the relative conservatism of the era of Leonid I. Brezhnev's leadership.

It is because of the ability of an older organizational generation not only to choose but to dominate the professional formation of new recruits that one may legitimately speak of the grip of the organizational past. Such a grasp is made up of the personnel and resource environment in which any innovation is attempted: change must use the people that the past has provided. If an entirely new cadre is imposed upon the organization, it has been destroyed, not reformed. The characteristics of personnel, the patterns according to which they change in an organization, and the degree of continuity, then, form a highly useful key for unlocking the mysteries of organizational structure. Thus, it is not only the fact of rapid change in personnel after 1917 that is important but its pattern. This pattern, manifesting itself in the 1920s, offers a useful means of understanding the new Soviet administration and of assessing its continuities with and departures from prerevolutionary structures.

Class, Career, and the Imperial Bureaucracy

On the basis of the foregoing observations, one could conclude that while bureaucracy grew in size in nineteenth-century Russia, it necessarily became less and less attached to its nobilitarian landed traditions. A common factor, offered to explain the rise of technocrats and the attenuation of the importance of the landed nobility is modernization: the rapid growth of population, urban centers, and manufacturing, together with the emancipation of the serfs and the more or less consistent Russian tendency to centralize economic and political power. A related conclusion would be that the growth of bureaucracy contributed to undermining the landed nobility in the late nineteenth century and that the nobility, for its part, had every reason to respond to this growing "alien" menace with hostility—not least because as the bureaucracy increased in size, it polluted the ranks of the landed nobility through the transformation of high-ranking bureaucrats into nobles.[30] These views may be taken as conventional wisdom received not only from the nineteenth-century nobility itself (the "defenders of privilege" in Seymour Becker's phrase) but from contemporary scholars such as Haimson, Hamburg, and Manning.[31]

Consideration of the role of the nobility as an estate or class in Russian bureaucracy requires the joint analysis of structural and demographic change together with bureaucratic recruitment and assignment patterns over time. As already noted, between 1801 and 1917 many changes were evident in important parts of the bureaucratic structure: the creation of ministerial government; the replacement of landlord-serf administration in 1861 with a comparatively formal, integrated system of regional and local civil administration; and the rapid expansion of staff and offices throughout the century. But although specialists multiplied, the Imperial bureaucracy did not support increasing specialization with the creation of specialized ministries or commissariats; specialists and generalists, made their careers within the confines of the same ministries. This circumstance might lead to ever gloomier assessments of the contemporary status and future chances of the nobility. "Under modern conditions," notes Reinhard Bendix, following Weber, "the only alternative to administration by officials who possess [expert] knowledge is administration by dilettantes. And this alterna-

---

30. Blum, *End of the Old Order*, 421–22.
31. Leopold Haimson, ed., *The Politics of Rural Russia, 1905–1914* (Bloomington, Ind., 1979), 7–8, 30–31, 262–63; G. M. Hamburg, "The Russian Nobility on the Eve of the 1905 Revolution," *Russian Review* 38 (1979), 323–38, and *Politics of the Russian Nobility, 1881–1905* (New Brunswick, N.J., 1984); Roberta Manning, *The Crisis of the Old Order in Russia* (Princeton, N.J., 1982).

tive is ruled out wherever the expert performance of administrative function is indispensable for the promotion of that order and welfare which is regarded as mandatory by the decision-making powers."[32]

The scenario is one of expansion, professionalization, and antitraditional bias against landed nobility playing itself out over the course of four generations. Toward the end of the nineteenth century, moreover, the entire scene acquired great vitality and menace for the "defenders of privilege" through industrialization and urbanization, elements creating a tidal wave of change that would sweep away the nobility as an estate, possibly as a social group, and surely as an independent socio-political force in Russian society. As Becker notes—correctly, I think—these findings are common not only to a significant body of Soviet and Western historians of nineteenth-century Russia but also to novelists, playwrights, and, broadly, spokespersons for the nobility at the time.[33]

The balance of this chapter extends the argument that the conventional wisdom about the role of the nobility is incorrect.[34] Briefly, I agree with Becker that, very far from declining, the role of the landed nobility in bureaucracy endured in the later nineteenth century. More important, I argue that nobilitarian success occurred in such a way as to leave a profound imprint on the structure of civil administration in Russia even beyond 1917. In particular, the role of the landed nobility was so closely identified with segments of provincial administration that the problems of land reform and administrative change were inseparable not only before but after 1917. As a result, although according to Stephen P. Sternheimer many segments of the bureaucracy rode out the storm of 1917 and crossed the revolutionary divide, I propose that provincial administration was much more likely engulfed in the same cataclysm that changed the structure of Russian landholding virtually overnight.[35]

This view, obviously, has important implications for my understanding of the course and outcome of the Revolution of 1917. Although Theda Skocpol is not alone in believing that the Russian peasantry played a pivotal role in the revolution in large measure because of the weakness of the nobility, in my view, she both misinterprets and under-

32. Reinhard Bendix, *Nation-Building and Citizenship*, rev. ed. (Berkeley, Calif., 1977), 196.
33. Becker, *Nobility and Privilege*, chap. 1.
34. See also Don Karl Rowney, "Organizational Change and Social Adaptation: The Pre-Revolutionary Ministry of Internal Affairs," in Pintner and Rowney, *Russian Officialdom*, 283–315, and "Structure, Class, and Career," *Social Science History* 6, no. 1 (1982), 87–110; Becker, *Nobility and Privilege*, chap. 1.
35. Stephen P. Sternheimer, "Administration for Development: The Emerging Bureaucratic Elite, 1920–1930," in Pintner and Rowney, *Russian Officialdom*, 316–54.

estimates the importance of the peasantry because she misunderstands the late nineteenth-century status of the nobility:

> Russia's czarist officialdom was renowned for its inefficiency and corruption, and yet it implemented basic agrarian reforms in 1861 and 1905 and administered the first stages of heavy industrialization.
>
> Leaving aside value-orientations and individual characteristics, we must look at the class interests and connections of state officials. *The adaptiveness of the earlier modernizing agrarian bureaucracies was significantly determined by the degree to which the upper and middle ranks of the state administrative bureaucracies were staffed by large landholders.* Only state machineries significantly differentiated from traditional landed upper classes could undertake modernizing reforms which almost invariably had to encroach upon the property or privileges of the landed upper class.[36]

Mine is a complex view: the nobility were weak in one way (because their landholding status was in eclipse) but vigorous in another (because they retained high status in parts of state administration and demonstrated their adaptability by continuing to manage a changing bureaucracy). In particular, this combination of economic vulnerability and adaptability on the part of the *dvorianstvo* focuses our attention on important differences between specific segments or sectors, of the prerevolutionary administration in Russia: the central and provincial, generalist and specialist. The following sections enlarge on this perspective.

The question of change in the conditions for induction into and advancement through the civil service can be considered independently on its own merits, but the problem of what persons were inducted into state service at various administrative levels and how they progressed during their careers is closely related to a bewildering array of other organizational characteristics. Organizational structure, moreover, is also closely related to the impact of social characteristics on the bureaucratic system, since inductees obviously brought at least some of their social characteristics with them into the service.

To understand the implications of these connections, one must extend the observations made earlier on the personnel complement, or *sostav*, of imperial administration, which expanded substantially throughout the nineteenth century.[37] Between 1796 and 1851 the general population increased from about 36 million to 69 million, or 91.7 percent, while the total of officials rose from 16,000 to 74,000 or about 360 percent. The increase in the number of officials was even more

---

36. Theda Skocpol, "France, Russia, China," *Comparative Studies in Society and History* 18 (1976), 185 (original emphasis).
37. Zaionchkovskii, *Pravitel'stvennyi apparat*, 221–22; Starr, *Decentralization*, 12–14.

dramatic between 1851 and 1903: from 74,000 to 385,000. The ratio of population to officials went from more than 2,000 to 1 in 1796 to about 300 to 1 in 1903.

Changes in who was inducted into service and who advanced were apparently related to the expansion of the service as well as to changes in its activities. The expansion of offices in the early nineteenth century resulted in some increase in the numbers of nobles who were not wealthy enough to own serfs and of non-nobles in the central and provincial administration.[38] The number of desirable posts to be filled simply exceeded the number of well-to-do nobles willing to fill them. The implications of this change are reflected in legislation of 1834 and 1856.[39]

In the early nineteenth century, inherited social criteria—usually in the form of legal estate designation—were the primary basis for career advancement.[40] That is, recruitment and advancement were legally hastened by noble class origin, a professional advantage literally ascribed to the officeholder not on induction but at birth. During the first half of the nineteenth century the legislation governing induction and promotion changed several times. In general, the import of the changes after 1834 was to lessen the advantages arising from inherited social class designations. By 1856 the law that continued to define entrance and advancement criteria until the end of the Empire was in place. The modified criteria were education, time in grade (seniority), and distinguished service. Inherited legal estate designations were no longer legally sanctioned, although class and estate designation continued to play an important role in determining access to education.

Were these legislative changes the result of a bureaucratic expansion and increased demand for officials so vast that nobility no longer had any practical meaning? Perhaps, to some extent. Pintner subjected his sample of 4,700 officials serving between 1846 and 1855 to multiple classification analysis, holding age constant.[41] He found that for this group, education was by far the best predictor of career success in both central and provincial agencies. It is important to note, as Pintner has, that these findings make the legislation of 1856 and even of 1834 rather after the fact. The mean age of the highest-ranked (ranks 1 to 5 of 14 ranks) officials in the 1846–55 group was fifty-two and that of the next

38. Pintner, "Civil Officialdom," 214.
39. Zaionchkovskii, *Pravitel'stvennyi apparat*, 33ff; PSZ, 1834, no. 7224, and 1856, no. 312377.
40. For an extended theoretical and philological discussion of the terminology of Russian social class structure and its interpretive implications, see Gregory L. Freeze, "The Soslovie (Estate) Paradigm and Russian Social History," *American Historical Review* 91, no. 1 (1986), 11–36.
41. Pintner, "Civil Officialdom," 237.

lower group (ranks 6 to 8) was forty-three. Even if we assume that they were all serving as late as 1855, those ranked highest would have completed their education in 1825, and the more junior officials would have completed theirs in 1834—in other words, before the legislation could be expected to have had any effect for most officials in the sample.

The motivation for the legislation, then, must have been, among other things, a desire to *bring the law into accord with practice*, and emphasis on the role of formal education was probably a function both of increasing government need for the products of higher education and—although Pintner disputes this—of the expanding presence of individuals who were disadvantaged by inherited social classification. Whatever the finer nuances, it is clear that the expansion of the bureaucratic structure in the early nineteenth century was accompanied both by a modest increase in the numbers of nontraditional social groups represented in the bureaucracy and also by substantial modification of recruitment and advancement criteria.

Even though social class, in spite of the legislation, remained an important factor for discriminating among broad segments of the bureaucracy, an important trend was established in the first half of the nineteenth century: nobility, especially landed nobility, were losing their position of automatic dominance in the civil service as a whole. Just how important this trend was may be seen in the implementation of the legislation on emancipation of the serfs.

Yaney notes that not only in passively accepting the emancipation and land reform measures but in actually participating in their administration, the nobility was acting against its own class interests.[42] I propose that we do not have to appeal to latent romanticism for an explanation of this behavior. It was not a case of noble landlords working against themselves to realize a Mallorian "myth of the nobility," as Yaney suggests. It was simply evidence of the fact that the real interests of an increasingly landless bureaucracy were to follow the orders they received from administrative masters who, although *they* may have held land, were sufficiently committed to the success of their careers and their policies to subordinate their identities as landholders. By midcentury Russian central government, perhaps by accident, had developed the landless administrative cadres that Trimberger and Skocpol identify as vital to the success of "revolution from above" in traditional societies.[43] Of 893 holders of top ranks (grades 1 to 4) in 1858, Korelin identified only 308 (about one-third) as possessors of inherited

---

42. Yaney, *Systematization*, 189–90.
43. Ellen Kay Trimberger, "A Theory of Elite Revolutions," *Studies in Comparative International Development* 7 (1972), 191–206; Skocpol, "France, Russia, China," 175–210, and *States and Social Revolutions* (New York, 1979), chaps. 2, 3.

land.⁴⁴ In the villages and towns, those members of the landed nobility who chose to work as administrators of the land reforms (for example, as mirovye posredniki) were probably doing so out of a desire to limit damage to their class or estate interests; but the lapidary fact remained that with them or without them, a civil administration in which the relative position of the noble estate was constantly eroding would implement the legislation.

Landholding and High Civil Office after Midcentury

The trend toward creation of a non-noble, landless civil administration continued throughout the remainder of the nineteenth century. However, an analysis of the recruitment and advancement of officials in the second half of the century reveals a particularly complex, changing relationship between the administrative system and the upper social classes in Russian society.

As we have already seen, the principal students of this subject, apart from Seymour Becker, have concluded that while traditional elites—the landed nobility in particular—maintained some of their traditional dominance in the senior administration, elsewhere in the bureaucracy their relative decline was evident. Specifically, some erosion in the relative position of the noble landholder (pomeshchik), with serfholding as the measure, has been observed by Pintner, who compares his findings of 1855 with those of Troitskii for the year 1755.⁴⁵ According to Zaionchkovskii and Korelin, moreover, the process was greatly accelerated in the second half of the century. Korelin finds that 53.5 percent of officials in grades 4 and higher held at least 100 desiatins (about 270 acres) in 1858, while data I have interpolated from Zaionchkovskii indicate that only 34 percent had similar holdings in 1878. By 1902 this figure had fallen, according to Korelin, to about 27 percent.⁴⁶

These relatively declining fortunes of the landed nobility in state service are summarized in Table 1, which shows also the increasing proportion of officials who held less than 100 desiatins of land. On the basis of such data Zaionchkovskii concluded that "toward the beginning of the twentieth century the number of pomeshchiki among all categories of the higher bureaucracy and provincial administration" had decreased.⁴⁷ It should be underscored, however, that both Korelin

---

44. A. P. Korelin, Dvorianstvo v poreformennoi Rossii: Moscow, Nauka, 1860–1904 (Moscow, 1979), 98, Table 8.
45. Pintner, "Civil Officialdom," 196–201; S. M. Troitskii, Russkii absoliutizm i dvorianstvo v XVIII v.: Formirovanie biurokratii (Moscow, 1974), 213–16.
46. Zaionchkovskii, Pravitel'stvennyi apparat, 90–105; Korelin, Dvorianstvo v poreformennoi Rossii, 98.
47. Zaionchkovskii, Pravitel'stvennyi apparat, 222.

TABLE 1.
Change in landholding among officials, ranks 1–4

Land holdings	1858	1878	1902
Under 100 desiatins	46%	65%	73%
100 desiatins or more	54%	34%	27%
N	893	2,509	3,642

Sources: Walter M. Pintner, "Civil Officialdom and the Nobility in the 1850s," in Pintner and Don Karl Rowney, eds., *Russian Officialdom* (Chapel Hill, N.C., 1980), 196–201; P. A. Zaionchkovskii, *Pravitel'stvennyi apparat samoderzhavnoi rossii v XIX v.* (Moscow, 1978), 90–105; A. P. Korelin, *Dvorianstvo v poreformennoi Rossii* (Moscow, 1979), 98.

and Zaionchkovskii are at some pains to indicate that the hold of the nobility in general—landholders among them—over the *highest* reaches of the administration was still strong up to the end of the Old Regime.

Unquestionably, the impact of the relative decline of landed nobility on the administrative system was considerable; certain aspects of this impact are reflected in the modification of laws on inherited social class. The nature of the changing relationship between upper-class landholding and officeholding, however, was not what the previous research suggests, and the reason is straightforward. While it is certainly true that the *proportion* of landholders in bureaucracy declined between 1858 and 1902, their *numbers* actually doubled among those holding the top four ranks: the figures for 1878 indicate a sharp increase—78 percent—in landholders over the 1858 data. The 1902 figures indicate a rise of about 13 percent over 1878, and of more that 100 percent over 1858.[48]

It is worthwhile to speculate on the meaning of these data. For the civil administration as a whole there was no evidence that the nobles could use their positions to limit or control administrative actions. During emancipation they cooperated in the implementation of the legislation. But what about the role that civil administration undoubtedly played as a source of employment, official status, income? Did the noble landholder who entered service in, say, 1878 perceive his chances of rising in the bureaucracy as better or worse than those of his father, uncle, or cousin who had entered twenty years before and who was now holding down a position in one of the top four ranks? Of course, the chances of any individual would be determined by a host of factors: ambition, money for education and social standing, intelligence, family connections, and so forth. Still, the class as a whole might be expected to perceive its possibilities at least partly on the basis of what had been

---

48. For an earlier formulation of the same argument, see Rowney, "Structure, Class, and Career," 98–99. Also see Becker, *Nobility and Privilege*, chap. 1, 108–11; Becker's figures are different, but his conclusions are similar.

achieved by others in the same social environment. Did the noble landholders have reason to see themselves as more and more disadvantaged by comparison to a landless service nobility or to non-nobles?

We can calculate the ratio of landholding top officials (*sanovniki*) to all landholders in forty-four provinces of European Russia at a given time. This is a kind of estimate of preferment for landholders if we accept the patterns of these provinces as a satisfactory indication of noble landholding throughout those areas in the Empire from which the service elites were drawn.

The figures are most interesting. In 1877 the ratio of landholders with more than 100 *desiatins* to *sanovniki* with at least that amount of land was 66 to 1. To me that indicates quite substantial levels of preferment: of every hundred or so landholders of substance, the odds were good that one or two were highly successful civil officials. Among many landed families, of course, there would have been more than one large landholder. And of course, in addition to the high *civil* officials considered here—governors, vice-governors, heads of departments, counselors to ministers, ministers—there were many high *army* officers among the nobility. The point, simply, is that in 1877–78 Russian landholders had very good access to high office, in spite of what the historical literature or the nobles' own recollections may have led one to believe.[49]

But I am more interested in the pattern of change over time—in particular between 1877 and 1902, a period during which the *rate* of increase in landholders occupying top positions declined. As we already know, the number of landholders in the general population had declined too, and it seems possible that this was symptomatic of an overall disintegration of the traditional position of influence enjoyed by the *pomeshchiki*. In fact, however, the ratio of landed *sanovniki* with more than 100 *desiatins* of land to landholders in general improved during this period by about 25 percent—from 1 in 66 to 1 in 50. Of course, limiting the sample of landholders to adult males of serving age—or even to landed families—would make the ratio even more favorable.

The implications of these data for the Russian bureaucracy are, to my mind, both interesting and highly significant. Whatever difficulties noble landholders faced in rural Russia, it seems hard to sustain the notion that they were losing their grip as a class on the higher reaches of the bureaucracy. In this tenacity they seem similar to the pre-nineteenth-century English landed elite as it is now seen by Lawrence and

---

49. Becker (*Nobility and Privilege*, 112–13) looks at *all* nobles and *all* landholders; he finds high but declining proportions from 1857 to 1897.

Jeanne F. Stone.[50] They were a social group quite capable, because of wealth and connections, of defending their privileges and of tightening their grip on social and political prerogative. Even as both law and commerce seemed to be undermining them, they were aggressively using the considerable resources at their command to enhance their opportunities.[51] This did not mean that the landed Russian nobility were able to halt the major state-building initiatives of the century; the government was far too powerful and nonlanded civil servants too numerous. Professionalization of bureaucracy, land reform, industrialization—all would go on with or without the support of the landed nobility. It did mean, however, that the nobility were in the best position to take personal and class advantage of the opportunities that law and commerce presented.

Noble dominance of the upper levels of the civil service and the general improvement within it of landholders' chances for career success do not, of course, tell the whole story of the relations between the traditional civil service class and the bureaucracy. There is a series of related problems, which can best be seen in terms of the location of specific groups of officials throughout the system. In this context, location means "geographical" with respect to the country or the society under study, but it also means "topographical" with respect to the organization. Much of the remainder of this chapter is concerned with both geographical and organizational location.

Organizational Status and Social Background

Let us take the problem of organizational location first. Earlier I referred to two dichotomously divided "sectors" of officialdom. Using a slightly different, nondichotomous approach, Armstrong proposed three: (1) professional-technical specialists, (2) provincials, and (3) bureaucratic generalists.[52] Whatever the logic of our classification, what we want to know is whether or not there are career and social differences between these sectors. In order to find this out, I focus on the first two categories and consider the bureaucratic generalists primarily in contrast to these two.

Were specialists and generalists likely to have come from the same or different social backgrounds or legal estates? This question obviously bears on both the pattern of Russian bureaucratic development and on

50. Lawrence Stone and Jeanne F. Stone, *An Open Elite? England, 1540–1880* (Oxford, 1984).
51. Rowney, "Organizational Change." For a review of the erosion of privilege in law, see Becker, *Nobility and Privilege,* 18–27; and Blum, *End of the Old Order,* 418–25.
52. Armstrong, *European Administrative Elite,* 226.

the capacity of a "beleaguered" nobility to retain its perquisites in a changing world. An examination of two populations of senior officials from two ministries—Ways of Communication; Trade and Industry—in the late nineteenth and early twentieth century helps to answer this question.

One group is described by data on the careers of fifth-rank and higher officials of the Ministry of Ways of Communication. This ministry trained a high proportion of its own officials in the Institute of Engineers of Ways of Communication, usually granting the successful candidates the degree of Engineer of Ways of Communication, an unmistakable mark of the specialist. This ministry was founded in 1865, at a moment when the technical segments of the civil administration were making their appearance everywhere.[53] If any organization in the Imperial civil administration provided an opportunity, justified by function, for complete control by technically trained "experts," it was the Ministry of Ways of Communication. Although its operational responsibilities had once included the post offices and the telegraph system, oversight of railway construction, and control of rural (zemstvo) roads, among other things, in the 1880s it was primarily identified with railway administration.[54] The identity was jealously guarded. The ministry refused to confirm the appointment of non-engineer Sergei Iul'evich Witte as manager of the southwestern railway system because, as Von Laue puts it, "according to the Ministry of Communications, where engineers had a monopoly, only engineers knew how to operate railroads."[55] Did engineers have a monopoly? If so, what were its social and organizational implications?

The career records of twenty-one department heads in the Ministry of Ways of Communication who served in the last few years before the revolution show seventeen (about 80 percent) from the hereditary nobility.[56] The remaining individuals were principally the children of officials and officers. The figure for nobility is about as high as one would expect for any group of higher civil servants—even higher, perhaps, since the overall average in the central administration was about 70 percent. Yet holders of 100 *desiatins* or more of land constituted only 10 percent of the ministry's higher officials. This contrasts with Korelin's figures of about 27 percent for all holders of 100 *desiatins* or

53. See p. 27 above.
54. N. P. Eroshkin, *Istoriia gosudarstvennykh uchrezhdenii*, 216–17.
55. Theodore H. Von Laue, *Sergei Witte and the Industrialization of Russia* (New York, 1969), 47.
56. Tsentral'nye gosudarstvennyi istoricheskii arkhiv Leningrada (TsGIAL), fond 229, op. 10; Ministry of Ways of Communication, *Spisok lichnogo sostava Ministerstva Putei Soobshcheniia: Tsentral'nye i mestnye uchrezhdeniia*, Izdanie kantseliarii Ministra (St. Petersburg, 1916).

more in ranks 1 to 4.[57] Whatever we may say about the landed minority, the overwhelming majority of these men were not from the Russian social elite.[58]

A comparison of the incumbents of the 1898 Engineering Council in the Ministry of Ways of Communication (thirty-two officials) with those of the Council of the Minister (seven officials plus the minister) results in some sharp contrasts. Of the full members of the Engineering Council, most were hereditary nobles, but none had received an elite education, and their landholdings were roughly comparable to those of the department heads just described. Their salaries tended to fall in the range of Rs 4,500 to Rs 5,500, perhaps about average for higher civil servants in the central administration but not munificent—especially if we consider that, on average, the group was well over fifty years old and might, on the grounds of seniority, have commanded more. As for the 1898 Council of the Minister of Ways of Communication, the minister's principal advisory body, all members were hereditary nobles and, significantly, no member was a product of the ministry's own institute. Instead, they hailed from such elite institutions as the Alexander Lyceum, the Imperial School of Legal Studies, or the Corps of Pages. Salaries were uniformly Rs 6,000 a year for this group whose members averaged nearly sixty years of age; that is, the compensation was only slightly better than that of members of the Engineering Council if one takes age or seniority into account. This comparison however, certainly does not suggest that a premium was placed on the professional contributions of engineers, whether or not they were in short supply in the general population. In addition, three of the seven members of the council were substantial landholders, two of them holding thousands of *desiatins*.

Does this mean that "engineers had a monopoly" or not? Certainly, on the one hand, they were in charge of the major operational sections— the departments—of the ministry, a dominance contrasting sharply with even the medical organizations within the Ministry of Internal Affairs, where specialists were distinctly on tap rather than on top.[59] On the other hand, these "technicians" clearly could not expect to rise to the highest political strata of the ministry. I would be prepared to

---

57. Korelin, *Dvorianstvo v poreformennoi Rossii*, Table 18.

58. I am indebted to Seymour Becker for calling my attention to the fact that possession of 100 *desiatins* of land at the end of the nineteenth century did not necessarily mark the owner as a member of the social elite (cf. Becker, *Nobility and Privilege*, 52–53). I continue to use the 100-*desiatin* category as a cutoff point because it is the one used by Korelin.

59. Don Karl Rowney, "Higher Civil Servants in the Russian Ministry of Internal Affairs: Some Demographic and Career Characteristics," *Slavic Review* 31, no. 1 (1972), 101–10; and "Structure, Class and Career."

call engineers influential and successful in these circumstances, but I do not think they had a monopoly. Von Laue missed another possible explanation for the ministry's hostility toward Witte: maybe the departments objected because he was not an engineer and the elite because he was not a bona fide member of the Russian upper crust.

Parallel comparisons can be made for the Ministry of Trade and Industry in 1912, another administrative locus where expertise might be expected to dominate. The available data here are not rich. Since information on social estate- and landholding are lacking, it is necessary draw inferences on the basis of educational background.[60] The Council of the Minister included sixteen members, of whom many were products of the elite Alexander Lyceum (including the minister himself). By contrast, the ministry's Technical Council and the Technical Assembly under the council were staffed with people possessing the credentials of engineers (for example, from the Institute of Engineers of Ways of Communication). Other departments, however, tended to be managed by officials who were products of the universities of St. Petersburg or Moscow, giving the impression that whatever the case in the Ministry of Ways of Communication, specialists definitely did not have a managerial monopoly here. They were important, but they were not in charge; their careers were truncated.

Additional data on the subject more clearly suggest relatively great non-noble and non-landholder presence in professional-technical agencies and the relatively subordinate status of the positions they occupied. As indicated above, the central apparatus of the Ministry of Internal Affairs consisted of agencies of widely diverse functions and professional characteristics. The largest block of agencies dealt with general administrative matters; others supervised police activities, censorship, medicine, veterinary affairs, civil engineering, and so forth.[61]

The organizational elite of the central ministry comprised some thirty-six officials between 1905 and 1916. Of these, twenty-eight were trained in general education establishments such as law schools or universities and were all employed in nonspecialist positions within the higher civil service of the ministry. Of the twenty-eight, eighteen (nearly two-thirds) spent substantial portions of their careers outside the capitals, St. Petersburg and Moscow; in the great majority of cases, sojourns outside the capitals were distributed throughout their careers. The provincial positions held by these officials were usually those of procurator, vice-governor, or even province governor. Of the twelve

---

60. Ministry of Trade, *Spisok lichnogo sostava tsentral'nykh uchrezhdenii Ministerstva Torgovli i Promyshlennosti k 1 sentiabria, 1912* (St. Petersburg, 1912).
61. Rowney, "Higher Civil Servants."

TABLE 2.
Ministry of Internal Affairs: Social origins (by division) of high officials, 1905

	Administration	Police	Information	Technical
Hereditary and foreign nobility	85%	100%	67%	61%
Non-nobles	15%	0%	33%	39%
N	70	5	13	13

Source: Ministry of Internal Affairs, Spisok vysshikh chinov Ministerstva Vnutrennikh Del na 1905 g. (St. Petersburg, 1905).

TABLE 3.
Ministry of Internal Affairs: Social origins (by division) of high officials, 1916

	Administration	Police	Information	Technical
Hereditary and foreign nobility	82%	67%	68%	62%
Nonnobles	18%	33%	22%	38%
N	79	9	18	16

Source: Ministry of Internal Affairs, Spisok vysshikh chinov Ministerstva Vnutrennikh Del na 1916 g. (St. Petersburg, 1916).

officials educated in professional-technical institutions or occupying technical positions within the ministry, however, all had spent at least the last half of their civilian careers in St. Petersburg. Unlike their generalist counterparts, of course, they tended to concentrate on functionally specific organizational activities.

Tables 2 and 3 summarize the social origins of officials in four employment categories in the Ministry of Internal Affairs in 1905 and 1916. Since these data include roughly similar ranks and generally equal offices in terms of grade and compensation (with the exception of a small number of the general administration offices), we may take age and—what is almost the same thing—organizational seniority to be constant. The proportions, then, are striking. The professional-technical agencies' officials were always somewhat less noble (more "middle class" in Armstrong's phrase) than the remainder of the officials. In addition to the class differences shown in Table 2, the data also indicate that professional-technical officials consistently either possessed no land or much less land than their colleagues in other divisions. Thus, Blum's statement strikes one as half true, not more: "Education and merit, instead of birth, became more and more the pass-key to high places. The always expanding scope of activity of the post-emancipation state created the need for more bureaucrats, and particularly for trained specialists. The nobility was neither large enough nor did it contain enough able men to meet the demand."[62]

Simple formulas are often suspect, human ingenuity in matters of precedence and status being what it is. Education and merit were indeed passkeys to better offices, but they unlocked only doors other than those in the building at No. 16 Fontanka Quay, where the ministerial elite had their offices. Entrance there required the right social credentials as well. Elsewhere in the civil bureaucracy, moreover, the inaccuracy of Blum's formulation is even clearer during the final twenty years of the Empire. In order to see this, let us consider the contrasts between the central and provincial sectors.

Differences between Provincial and Central Sectors

Pintner has shown conclusively that in the first half of the nineteenth century there were manifold differences between provincial and central officials; provincial officials were less well educated and tended to hold inferior positions at inferior pay. In considering higher ranks only, however, he finds some crossing back and forth between provincial and central offices—as, indeed, I have done in the data cited above for generalists in the Ministry of Internal Affairs.[63]

Another look at service employment patterns for higher provincial officials is helpful. Pintner found that in midcentury (1840–55) about 42 percent of the provincial elites were employed initially in the military. Of sixty-nine governors who held office in 1916 (virtually all of whom were first appointed governor before World War I), I found that 50 percent began their careers in the military. Moreover about a fifth of the 1916 governors began their careers in the central administration as compared to 28 percent who began in the provinces. Comparable figures for Pintner's group were 30 percent central and 17 percent provincial. These are not large differences; in other words, when one compares mid-nineteenth-century and 1916 senior provincial careers, the similarities seem more impressive than do the differences.

Going further, as Table 4 shows, thirty-nine of sixty-nine governors (56.5 percent) had served as vice-governor immediately before their first appointments as governor. Of the remainder, the large majority had served in other territorial positions. Of course, some or most of

62. Blum, *End of the Old Order*, 421.
63. Pintner, "Civil Officialdom"; Richard G. Robbins, Jr., "Choosing the Russian Governors: The Professionalization of the Gubernatorial Corps," *Slavonic and East European Review* 58, no. 4 (1980), 546–58, Table 3; G. M. Hamburg, "Portrait of an Elite: Russian Marshals of the Nobility, 1861–1917," *Slavic Review* 40, no. 4 (1981), 585–602, esp. 597ff.

TABLE 4.
Immediately prior positions held by provincial governors, c. 1891–1916

	Governor (entered)	Vice-governor	Civil administrator	Petersburg (Internal Affairs)	Petersburg (other)	Elected roles	Other
N	10	39	7	4	1	4	4
%	14.5	56.5	10.1	5.8	1.4	5.8	5.8

Source: Ministry of Internal Affairs, *Spiski vysshikh chinov Ministerstva Vnutrennikh del* (Moscow, 1896–1916).
Note: No governors were previously land captains or members of military governments.

Domestic Administration before 1917

TABLE 5.
Immediately prior positions held by provincial vice-governors, c. 1891–1916

	Vice-governor (entered)	Elected roles	Civil administrator	Land captain	Other
N	9	9	17	1	5
%	23.1	23.1	43.6	2.6	7.7

Source: Ministry of Internal Affairs, Spiski Vysshikh Chinov Ministerstva Vnutrennikh Del (Moscow, 1896–1916).

those whose first provincial appointment was that of governor could have come from St. Petersburg, the only likely alternative being the military; even so, altogether they accounted for less than 15 percent of the total cohort. Finally, of the thirty-nine governors who had previously been vice-governors, thirty (77 percent) had held other *provincial* posts before securing the vice-gubernatorial appointment. Again, it is quite possible that the nine whose first provincial appointment was vice-governor (see Table 5) came directly from St. Petersburg, but it should be noted that these officials were outnumbered decisively by individuals who held civil administrative posts in the provinces (seventeen) and equaled by those who held the rather more unusual elected positions such as district or provincial marshal of nobility.

The last cohort of senior provincial bureaucrats, then, whose average date of appointment was about 1913, had behind them careers that separated them from their institutional peers in St. Petersburg. The most distinctive features of their career patterns were extended service in the provinces and a tendency to have begun in the military or in some provincial post. The comparison with Pintner's data, moreover, suggests that for landed nobles this tendency, like access to higher civil service itself, was at least holding firm—if not improving to the advantage of the landed nobility—during the last two generations of the monarchy. Emphasizing the singularity of this career pattern, mid- to late-career appointments were also overwhelmingly provincial: civil administrator, vice-governor, governor. In order to avoid a facile conclusion that links these careers to a nobilitarian resurgence following the Revolution of 1905, moreover, it should be noted that many of the career events—especially midcareer appointments—occurred before the Revolution of 1905.

These successful provincial careers were connected to the easy access of landed families to both good education and high-level administrative appointments. To demonstrate this proposition requires an examination of provincial administration between 1896 and 1916 from several points of view.

First, comparing the social characteristics of the senior provincial administration with those of the highest four ranks in the bureaucracy overall makes the contrast clear: the provincial administrative elite was much more noble and had more land. Zaionchkovskii's findings for the landholdings of governors in 1903 can be compared with Korelin's findings for the top four ranks in 1902. As noted above, Korelin found that landholders accounted for 26.6 percent of the total of 3,542 officials, but Zaionchkovskii found that more than two-thirds of the forty-eight governors he studied held 100 *desiatins* of land or more, and all governors for whom he had data were hereditary nobles; in his sample of vice-governors, 93.7 percent were hereditary nobles and 60 percent were landholders. These figures seem to confirm an exceptionally strong interest in high provincial office among the landed nobility and consistent success in attaining it; they also underscore the contrast between provincial and central office holders, a contrast that sharpened during the last seventy years of the empire.

The Importance of Career to the Nobility

The continuing success of landed nobility in achieving high bureaucratic office, far from demonstrating decline, would appear to highlight their power and their ability to use that power when their status and economic interests were at stake. The simplest measure of this enduring class vitality is certainly the fact that even as its *proportions* declined in a growing bureaucracy, the nobility's *numbers* increased substantially—not merely in the organization as a whole but at its highest, most influential levels. However, it is important to be clear about what power means here. Access to desirable offices does imply power over organizational events such as hiring, appointment to specific posts, and the allocation of resources within the organization; however, it need not— and, in the Russian case, did not—imply power over whether major policies would be implemented and how this would happen. Even by midcentury, as we have seen, the civil service was well beyond the controlling grip of the nobility. Thus, at one and the same time, bureaucracy could serve the career interests of the nobility and of ambitious non-nobles, and the state-building interests of political leaders.

One important implication of these findings is that the landed Russian nobility experienced a change in career ambition or motivation as a class in the later part of the nineteenth century,[64] the most eloquent testimony to which is the changing ratio of landed high officials to

---

64. Becker, *Nobility and Privilege*, chap. 6; Korelin, *Dvorianstvo v poreformennoi Rossii*, chap. 2.

landed nobility generally, discussed above. Some students of this problem, misled by the changing ratio of landed nobility to officials generally, have concluded (as Zaionchkovskii did) that the numbers of nobility in civil administration decreased. They did not. Still others have dismissed these conclusions, by implication, as spurious. Blum, for example, seems to suggest that the demographic flow was in the opposite direction, that the significant trend was for low-born bureaucrats to achieve high office and thereby a patent of nobility.[65] Such a trend would both dilute the "real" nobility and cast their role in the bureaucracy in an ever darkening shadow.

In Russia, however, ennoblement of either army officers or career bureaucrats was not easy in the second half of the nineteenth century (it had generally become more difficult throughout the century), and accession even to civil rank 5 (state counselor) conferred only personal, or life, nobility.[66] For these and other reasons a significant addition to the ranks of hereditary nobility via the civil bureaucracy is highly doubtful. Although Blum correctly observes that the Russian hereditary nobility expanded enormously in the second half of the nineteenth century, the general population expanded even more rapidly. In 1897 the hereditary nobility had increased to 148 percent of its 1858 size, but in roughly the same geographic area the general population had expanded to 177 percent of its 1851 figure.[67] Most of the expansion of the hereditary nobility as a class is probably explained not by an influx of ennobled commoners but by the same factors that account for the expansion of the general Russian population: improved nutrition and access to health care, and improved protection from communicable disease in the country as a whole. Moreover, if we consider land held by officials in terms of its source—perhaps the best indication of class origins, independent of the legal designation that appeared on each official's career dossier—we find that the proportion of inherited land to land acquired by all other means hardly changed between 1858 and 1902. As Table 6 shows, the proportions for 1858 and 1902 are less that 3 percent apart.

In addition to the foregoing reasons for thinking that ennoblement of bureaucrats did not have much impact, one can estimate—using a combination of estate censuses and military and civil service censuses—the numbers of officers and *chinovniki* who were promoted high enough to merit ennoblement in the second half of the nineteenth century. Altogether, I estimate the maximum total addition to the hered-

65. Blum, *End of the Old Order*, 422.
66. Eroshkin, *Istoriia gosudarstvennykh uchrezhdenii*, 313–14.
67. Korelin, *Dvorianstvo v poreformennoi Rossii*, 292–96; A. G. Rashin, *Naselenie Rossii za 100 let (1811–1913 gg.): Statisticheskie ocherki* (Moscow, 1956), 44–46.

TABLE 6.
Sources of land (100+ desiatins) held by top imperial civil servants

	Inherited	Purchased	Grants	Unknown
1858				
N	308	132	35	4
%	64.3	27.6	7.3	0.8
1902				
N	597	325	19	24
%	61.9	33.7	2.0	2.5

Source: A. P. Korelin, *Dvorianstvo v poreformennoi Rossii*, (Moscow, 1979), Table 8.

itary nobility from the ranks of commoner civil servants and army officers at less than 10 percent from 1861 to 1917. Using different and generally better data, Becker reaches similar conclusions: "Even if a growing proportion of unreported cases is assumed for the period after 1858... the relative weight of the newcomers in 1897 would hardly have exceeded 8 percent."[68]

This does not mean that successful commoner bureaucrats did not thirst after and attain patents of nobility, even hereditary nobility, in the waning years of the Empire. Nor does it mean that these same persons did not acquire both land and the other material attributes of traditional noble status. Successful non-noble bureaucrats secured both, right up to the end. What is questionable is whether they did so in numbers large enough to have a meaningful impact on what is meant by "hereditary nobility" in Russian society and in the highest reaches of the bureaucracy. In my judgment they did not, and thus I argue that the data describing high-level bureaucratic officeholders are really telling us about the careers of Russia's traditional social elite: the hereditary, landed nobility.

Bureaucratization in the Provinces

Were the nobility a cause or merely the beneficiaries of expansion? As time passed following the emancipation of Russian peasants, more and more of the landed nobility sought out bureaucratic careers. Why? Were these positions simply a substitute for the increasingly rare landlord status of their fathers? The answer has to be sought as part of the response to a larger question. What factors drove the expansion of bureaucracy after emancipation? I have already answered this question in a general way: emancipation implied a whole host of new administrative responsibilities. But as S. F. Starr notes, bureaucratic expan-

68. Becker, *Nobility and Privilege*, 90ff.

sion, especially in the provinces, had proceeded apace under Nicholas I—in the 1830s and 1840s.[69] It thus seems necessary to consider expansion, especially provincial expansion, independently of both emancipation and the interests of the nobility. Doing so means evaluating the most probable explanations by looking at the pattern of provincial bureaucratic structure all across European Russia for an extended period of time.

What are the most likely explanations? If we were to follow the direction in which new responsibilities after emancipation logically pointed, we might reach this conclusion: in Russia, bureaucracy replaced the traditional landlord-serf relationship in the countryside, making Russian bureaucratization primarily a rural phenomenon. As the independent roles of peasant communes increased and the size of the population doubled, and doubled again, the government responded by expanding its rural administrative cadres and institutions. This explanation would suggest an effort to make up in some measure for the "neglect" of the countryside before 1860. It would also explain continued noble dominance of the countryside and even of the bureaucratic elite. For while in their initial phase the postemancipation reforms were in the spirit of decentralization and self-government, they quickly reverted to the normal-for-Russia centralizing, bureaucratic tradition.[70] Moreover, even in its "self-government" phase, postemancipation administration was safely under the control of the former serf owners. Whatever the nuances, landlords and ex-landlords went to work at what they knew best—ordering peasants around. The systems created for local self-government—the zemstvos and their local administrations, the Main Administration for Affairs of Local Economy, and other local structures would all be part of an administrative network that would continue noble control of provinces. As such, this administrative network would be cut from the same coarse fabric first woven in the looms of Nicholas I, the gendarme not merely of Russia but of all Europe.

An alternative hypothesis gives a somewhat different view of things, however. If bureaucratization, although not totally unrelated to events in the countryside, should prove to have been mainly a phenomenon-related commercial-industrial-urban growth, then the explanation for continued noble dominance of the bureaucratic elite would be somewhat different. This hypothesis might also change our perspective on Russian state-building. If bureaucratization was linked to modernization not only in the 1890s, the era of Russia's industrializing spurt, but also before emancipation in the 1860s, then it must be seen both as a

69. Starr, Decentralization, 12ff.
70. Ibid., 292–345.

symbol of central government's preoccupation with development and as one of the major implements in that process of development. Instead of being merely an effort to control and exploit the peasantry, mid-nineteenth-century bureaucratization would mainly be linked to urbanization, commercial development, and early industrialization. In any case it would be more obviously linked to modernization than to the "rural control" objectives outlined above.

In order to decide which of these interpretations is the better one, one can examine the distribution of low-level members of the official estate (*chinovniki* or personal nobles) in the European provinces of the Empire at the moment of emancipation (approximately 1860) and one generation later (about 1897) and infer from these data the presence of upper-level administrators.[71] These are demographic data: the number of males enumerated as *chinovnik* and personal nobles; as such they are by no means an exact measure of bureaucratization or even of the presence of bureaucrats, but they will serve.

Further, one can gauge the pattern of distribution of officials across forty-two provinces, province by province, by evaluating the relation of each to several other measures. These include three groups of variables: the first measures population characteristics generally; the second, traditional power of the landed nobility; and the third, modernization (see Tables 7 and 8). Of the fifty provinces of European Russia, seven (Lifliand, Kurliand, Estland, Astrakhan, Don Oblast, Bessarabia, and Arkhangelsk) must be excluded for lack of adequate data. St. Petersburg province is also excluded because of its special and obviously abnormal status: as the seat of central government and a magnet for elites, it would inevitably distort findings for other provinces.

Both the 1860 and 1897 structures are evaluated by a multiple regression analysis that reduces the total of seven or nine "candidates" for inclusion as independent, explanatory variables to three and two in 1860 and 1897 respectively (see Figures 1 and 2). Most striking is the tendency for "urban population" to be strongly associated with the presence of officials not only in 1897 but, although much more weakly, in 1860 as well. This association implies far more than the simplistic idea that *chinovniki* and personal nobles lived in towns. Urban population is a far more powerful explanation of bureaucratization than "population density," for example, even though the two are closely linked in the bivariate correlations (Table 7). Moreover, urban population is linked in the analysis of both 1860 and 1897 patterns to other indications of modernization: relatively low mortality rates (1860) and

71. Becker, *Nobility and Privilege*, 95–99.

TABLE 7.
Distribution of officials correlated with demographic data, 1860 (excluding St. Petersburg)

	I	II	III	IV	V	VI	VII
Populations							
I. Chinovniki	1.00						
II. Total population	0.36						
III. Population density	0.34	0.38					
IV. Noble population	0.06	−0.19	0.03				
Land							
V. Noble land	0.09	0.07	−0.37	0.23			
VI. Proportion of land held by nobility	0.19	−0.26	0.25	0.59	0.47		
Modernization							
VII. Urban population	0.48	0.24	0.55	−0.02	0.02	0.23	
VIII. Mortality rate	−0.14	0.25	0.02	−0.66	−0.16	−0.53	0.19

Sources: (I) A. P., Korelin, *Dvorianstvo v poreformennoi Rossii* (Moscow, 1979), app. 2, 298–303 (data for 1858); (II) A. G. Rashin, *Naselenie Rossii za 100 let (1811–1913 gg.): Statisticheskie ocherki* (Moscow, 1956), Table 19, 44ff. (data for 1863); (III) Rashin, *Naselenie Rossii*, Table 45, 77ff. (data for 1863); (IV) Korelin, *Dvorianstvo v poreformennoi Rossii*, app. 1, 292–97 (data for 1858); (V) Ministry of Finances, *Tsifrovye dannye o pozemel'noi sobstvennosti v Evropeiskoi Rossii* (St. Petersburg, 1897), Tables 3–4, 22–33 (data for 1859); (VI) ibid.; (VII) Calculated from Thomas S. Fedor, *Patterns of Urban Growth in the Russian Empire during the Nineteenth Century* (Chicago, 1975), app. I, 183ff.; (VIII) Rashin, *Naselenie Rossii*, Table 145, 187–88 (data for 1861–65).

TABLE 8.
Distribution of officials correlated with demographic data, 1897 (excluding St. Petersburg)

	I	II	III	IV	V	VI	VII	VIII	IX
Populations									
I. Chinovniki	1.00								
II. Total population	0.50								
III. Population density	0.58	0.33							
IV. Noble population	0.27	0.11	0.35						
Land									
V. Noble land	0.16	0.35	−0.13	0.19					
VI. Proportion of land held by nobility	0.17	0.11	0.52	0.70	0.40				
Modernization									
VII. Urban population	0.89	0.36	0.53	0.24	0.05	0.15			
VIII. Manufacturing workers	0.79	0.21	0.46	0.18	0.02	−0.02	0.77		
IX. Mortality rate	−0.19	0.12	−0.21	−0.59	0.07	−0.45	−0.14	0.02	
X. Proportion literate	0.27	−0.29	0.03	0.40	−0.12	0.02	0.36	0.47	−0.34

Sources: (I–IV, VI–VII) see Table 7 (data for 1897); (V) see Table 7 (data for 1896); (VIII) Fedor, *Patterns of Urban Growth*, Table 23, 140–42; (IX) Rashin, *Naselenie Rossii*, Table 145, 187–88 (data for 1897); (X) ibid., Table 257, 308–8.

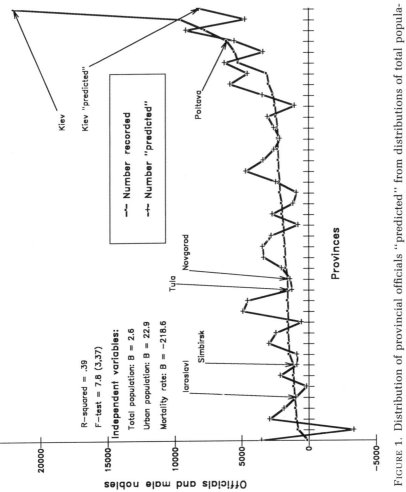

FIGURE 1. Distribution of provincial officials "predicted" from distributions of total population, urban population, and mortality rate, c. 1860 (excludes St. Petersburg and eight other provinces). Sources: See Table 7.

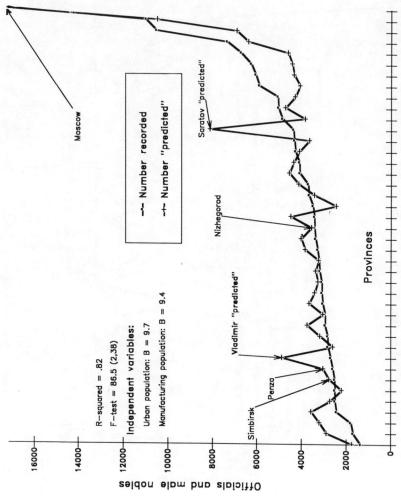

FIGURE 2. Distribution of provincial officials "predicted" from distributions of urban

Domestic Administration before 1917

high concentrations of populations engaged in manufacturing (1897), both of which explain a substantial amount of the variance when other factors are controlled. Even in 1860, concentrations of nobility are not closely associated with concentrations of officials. And although the bivariate (zero-order) correlations between noble population and factors such as *proportion* of land held are interesting, the nobility as a factor overall in bureaucratization is not sufficiently important to serve as the foundation for an argument that bureaucracy was a kind of indoor relief for nobles, expanding where their concentrations were greatest.

Striking also is the overall similarity between the included explanatory variables in the regression equations for 1860 and for 1897. Unfortunately, there is no measure for "manufacturing workers" for 1860. But I suspect that if there were, it would replace "mortality rate" as the third variable in the equation: mortality, negatively related to bureaucratization, is another measure of modernization.

Thus it is difficult to escape the conclusion that to some degree bureaucratization was driven by similar factors before and after emancipation. Although the relationship is far more solid in the 1890s (with a total explained variance of 87 percent) than a generation earlier (with an explained variance of only 39 percent), the included variables seem sound, historically relevant, logical, and parallel.

The issue of historical relevance is particularly helpful here. What kind of corroborative evidence is there that even before 1860 it was commercial-industrial-urban interests that were driving a significant share of bureaucratic expansion? It comes mainly from the legislation on the formation of local agencies in the major ministries: Internal Affairs, Finances, Justice, and Education. The extension of some of these ministries into the provinces with agencies that were linked to development and commerce began as early as 1803 (Ministry of Education) and 1809 (Main Administration of Ways of Communication). But it started in earnest under Nicholas I in the 1830s with the creation of permanent provincial treasury boards, customs offices, mine and salt administrations, and manufacturing and commercial divisions in the Ministry of Finances.[72] One could argue, moreover, that with landlords responsible for much of the behavior of serfs, the extension of police and judicial authorities into the provinces and localities was meant to serve the needs of the free, urban estates.

The two regression equations illustrated in Figures 1 and 2 describe a "normal" or average relationship among their variables. It can be visualized concretely by identifying "cases" (provinces) that were closely modeled or "predicted" by the equations. This is clear from an

---

72. Eroshkin, *Istoriia gosudarstvennykh uchrezhdeniik*, 180–82.

examination of the equation "predictions" themselves as well as the residuals or errors of the equations which result from comparing the predictions with actual values. The smaller the aggregate of residuals, the more accurately the statistical or theoretical model accounts for it.

In the 1860 analysis, provinces that were closely identified with the combination of prediction variables (total population, urban population, mortality rate) included Simbirsk, Tula, Iaroslavl, Novgorod, and Poltava (see Figure 1). Poltava, for instance, had one of the most rapidly growing urban populations in the early nineteenth century.[73] Its growth in this respect exceeded that of Kiev, for example, which the equation badly misidentifies. The reason is straightforward: for its level of urbanization, Kiev province had a much larger than normal official population.

Typical of the 1897 group were Moscow, Nizhni Novgorod, Simbirsk, and Penza (see Figure 2). By this time the full force of industrialization was being felt in the central provinces of European Russia; thus the typical or normal provinces identified by the equation are concentrated there. Moscow was the only province whose urban population exceeded the national average by a wide margin. The others ranged from slightly less than 100,000 to slightly under 200,000. The point is simply that urbanization *in combination* with industrialization created the conditions that explain a concentration of officials. Saratov and Vladimir provinces are substantially over- or underpredicted by the equation, but again it is Kiev, with its abnormally large population of officials, that stands out most clearly.

For provinces overall, not only at the end of the century after industrialization had begun but already in midcentury before emancipation, modernization and bureaucratic development were closely linked. Or, to put the case more accurately, bureaucracy was well and long established as the implement in Russia for controlling urban, commercial, and industrial development. It was the quintessential Russian framework of modernization, equaling at least, if not surpassing, the role of markets in the West in its importance for development. Not only was this characteristic important during the first spurt of industrialization in the 1880s and 1890s, but it had already been powerful in the 1850s, and, as we shall see, it created the inevitable and indispensable framework for development after 1917.

This analysis would seem to eliminate both the nobility and the peasantry as causes of bureaucratization. Their interests, or the need to control them either before or after emancipation, or before or after industrialization, would not seem to have been powerful and pro-

---

73. Rashin, *Naselenie Rossii*, 90.

nounced enough to determine patterns of distribution of offices or officials in the provinces. Moreover, as in the civil service overall—provincial and central—the nobility lost its legally privileged access to office, if not its access to privilege, as early as the 1820s and 1830s. Nevertheless, noble access to high office endured even, in the central government, to the point of blocking access for non-noble specialists—men of talent, education, experience, and (important for our purposes) expertise.

Regional Variation and the Power of the Nobility

What does this "urbanization" finding mean for the provincial sector of the Imperial bureaucracy?

Becker, after careful analysis, found that the proportion of all nobles (landowning or not) who were civil servants or officers declined in the second half of the nineteenth century, although their numbers rose.[74] He argues that continued noble interest in the civil service was the consequence of declining traditional roles, including landowning. If this is the correct explanation, then it implies that the bureaucracy's gain was landowning's loss. I propose that this is true up to a point but that there is an important sense in which the noble connection with the land remained and played a major role in the last act, the destruction, of the noble bureaucracy. In addition to accepting the hypothesis that declining landowning explains noble interest in non-estate careers and professions, I am also arguing that high-level officeholding in provincial Russia, even as it responded to modernization, continued to be open to the rural landed nobility. The nobles used their land—the connections and the status it conferred—as a means of securing bureaucratic preferment right up to the end of the Old Regime. This can be demonstrated by the patterns of variation in provincial elite appointments.

On its face, the idea that there was significant regional variation in officeholding when appointments were controlled from the center seems difficult to accept. Homogeneity, increasing across time, seems more logical. Uniform standards of admission and advancement, and uniform organizational objectives—in two words, professionalization and cosmopolitanism—should all certainly work against local or regional differences of every kind, especially in an autocracy. Moreover, we have a right to assume that with advancing modernization even the provincial/central sectoral variation evident in the early part of the nineteenth century would tend to disappear. Yet many students of rural

74. Becker, *Nobility and Privilege*, 108–16.

Russian society have noted significant variation in such diverse characteristics as farming methods, degrees of modernization, income from the land, and investment.[75] Variation in degree of industrialization was already a topic of intense interest in the nineteenth century.[76] Thus, the picture is far from clear, even when one has a fairly full sense of the pattern of distribution of officials and offices over many years. One must look at specific cases of officeholding in the provinces just as we have already done in the central government. What we find is that the "modernization" link affected not only the general pattern of office distribution but the bureaucrats themselves. That is, high officials in relatively urban, literate, industrial provinces were different from those in predominantly agricultural or traditional provinces.

The evidence consists of the details of post assignments, the career and biographic characteristics of the last generation of governors and vice-governors of thirty-five provinces where the land and family connections of the noble social elite were most indisputably present.[77] They extended from Petersburg in the Northwest to Samara in the Southeast; from Chernigov on the Left Bank of the Dnieper River to Perm and Orenburg in the East; they included Vologda and Olonets in the North and ran to Tavrida and Ekaterinoslav in the South. (Excluded are Astrakhan, Arkhangelsk, Bessarabia, the Don *Oblast* and Ufa—for want of data—as well as the non-Russian Lifliand, Estland, Kurliand, the Right-Bank Ukraine, Minsk, Mogilev, and of course the Polish provinces.) In all, they form a marvelous cacophony of language, religion, ethnicity, and even geography. Were it not for the uniformity apparently imposed by the Imperial army and civil service, we might expect nothing but variation, "deviance," from the urban Russian norm.

The evidence argues that governors possessing land, the mark of high social status, were assigned to posts in the "best" provinces. Moreover, the data argue that while simple possession of land was helpful in securing a good appointment, more land was better.

The patterns in Table 9 show that the segment of the final generation of tsarist governors (and perhaps vice-governors) who owned land received appointments in the capital provinces of Moscow and St. Petersburg and other relatively urban provinces; they were assigned to

75. Geroid T. Robinson, *Rural Russia under the Old Regime* (New York, 1932); Ainsley Coale et al., *Human Fertility in Russia since the Nineteenth Century* (Princeton, N.J., 1979); Barbara Anderson, *Internal Migration during Modernization in Late Nineteenth-Century Russia* (Princeton, N.J., 1980); Ivan D. Koval'chenko and L. V. Milov, *Vserossiiskii agrarnyi rynok XVIII-nachalo XX veka: Opyt kolichestvennogo analiza* (Moscow, 1974).
76. M. Tugan-Baranovskii, *Russkaia fabrika v proshlom i nastoiashchem*, 2 vols, 7th ed. (Moscow, 1938).
77. Ministry of Internal Affairs, *Spisok vysshikh chinov Ministerstva Vnutrennykh Del. Chast' II* (St. Petersberg, 1914, 1916).

TABLE 9.
Population characteristics of 35 provinces, 1916 (mean values)

	Governors' land (desiatins)	Vice-governors' land (desiatins)	Proportion urban (%)	Persons per sq. verst	Proportion literate (%)	Noble males (N)
Capital provinces (2)[a]	3,657.5	1,494.5	63.4	101.4	47.7	30,750.5
Other provinces with landed governors (20)[b]	1,444.8	1,149.2	11.7	50.2	21.1	5,279.4
Provinces with landless governors (13)[c]	0	1,125.8	9.8	42.5	19.8	3,436.1

Sources: Ministry of Internal Affairs, *Spisok vysshikh chinov Ministerstva Vnutrennikh Del na 1916 g.* (St. Petersburg, 1916), pt. 2 (land); A. G. Rashin, *Naselenie Rossii za 100 let* (Moscow, 1956), Table 57 (urban data for 1914), Table 45 (density data for 1914), and Table 257 (literacy data for 1897); A. P. Korelin, *Dvorianstvo v poreformennoi Rossii* (Moscow, 1979), app. 1 (noble males).

[a]Moscow, St. Petersburg.

[b]Iaroslavl, Kazan, Kharkov, Kherson, Mogilev, Poltava, Pskov, Riazan, Samarkand, Saratov, Smolensk, Tambov, Tavrida, Tula, Tver, Viatka, Vitebsk, Vladimir, Vologda, Voronezh.

[c]Chernigov, Ekaterinoslav, Kaluga, Kostroma, Kursk, Nizhegorod, Novgorod, Olonets, Orenburg, Orlov, Penza, Perm, Simbirsk.

the most literate provinces and those where there was a relatively high concentration of nobility. The implied advantage is simply that in such provinces there would be a greater opportunity for the development of contacts useful in furthering their own careers and the careers of their clients, relatives, or favored colleagues. By implication, too, I am saying that officials with substantial landholdings and other marks of superior social status tended to escape appointment to relatively rural, backward provinces. The civil servants who spent large portions of their careers there were a rather different breed.

A few examples will illustrate the differences that statistics present synthetically. Nikolai Antoninovich Kniazhevich was appointed governor of Tavrida province in the autumn of 1914. A scion of one of the hereditary noble families, he held some 2,500 *desiatins* of land, also in Tavrida—the coincidence was not unheard of, but it was unusual. He had attended the elite Alexander Lyceum and afterward the First Pavlovskii Military School. Kniazhevich's early career assignments were all in the elite regiment of Light Guards Hussars. After receiving his lieutenancy at the age of twenty-five, he rose within four years to the position of regimental adjutant. The next year, 1901, he was appointed commander of His Majesty's squadron in the regiment. Kniazhevich became deputy regimental commander in 1907 and, in 1912 at the age of forty-two, a major general attached to His Majesty's suite. His appointment as governor of Tavrida, a relatively literate province where the Russian nobility still held substantial lands and (not incidentally) the site of the vacation spa of the royal family, could hardly have been a surprise.

Another hereditary nobleman, Count Nikolai Vladimirovich Kleinmikhel, came to his first elite provincial post by a different route. Six years younger than Kniazhevich, he also attended the Alexander Lyceum but did not enroll in a military service school immediately on graduation. Instead, he managed an appointment to His Majesty's Horse Guards, a position he held only briefly before beginning a career in what could only be called rural noble politics: he became a justice of the peace, district marshal of nobility, and, at the age of forty, marshal of the nobility of Kharkov province. Count Kleinmikhel held 889 *desiatins* of land, a fact that doubtless stood him in good stead while seeking election to positions in the various noble assemblies; it was unusual for a marshal of nobility to be without land. In 1916 his promising career was given official approbation with his appointment, at the age of thirty-nine, to the vice-governor's post of Moscow province. Moscow, of course, was exceptionally urban, industrial, and literate by comparison with the vast majority of provinces.

These individuals and the others like them who held the most desirable elite positions in provincial administration were hardly specialists of any conceivable sort—unless one wishes to insist that career-building is a specialization. The point, however, is that the links between industrialization and urbanization on the one hand and Russian bureaucratization on the other did not imply that specialists—persons with experience in industrial management, or even persons with experience in bureaucratic agencies where economics and the collection and interpretation of statistics on trade and development were important—would receive the commanding appointments. On the contrary, these links meant that because these were appointments to the most comfortable, interesting, and promising places, the nobility with power and influence would assure that such places were reserved to themselves.

A final illustration may suffice. Olonets was among the least densely populated, least urban, least industrial provinces in European Russia. Although, because of its proximity to Petersburg province, it was not among the least literate—about average in this respect—its mortality rate was one of the highest in the Russian Empire. The last tsarist governor there, Mikhail Ivanovich Zubovskii, was appointed in 1913. Although it was Zubovskii's first appointment to a gubernatorial post, he did not receive the position in Olonets until his forty-ninth year, his twenty-fourth in service.

Zubovskii had attended the University of St. Petersburg, receiving a good but not an elite education. Noteworthy in the context of earlier discussions in this chapter is the fact that Zubovskii was in no sense a technical specialist. Neither did he have land or other marks of high social status—such as a title, or a wife from the social elite. Zubovskii's career was certainly successful, but it was sharply different from the two summarized above. He spent his early years in the judicial administration instead of in the military or in noble politics. In 1906 at the age of forty-two he became vice-director of the Department of Police in the Ministry of Internal Affairs—a position often reserved for jurists. Six years later he was appointed director of the chancellery of the Ministry of Internal Affairs and, eighteen months after that, governor of Olonets, where he was the top official until 1917.

The dominance by the landed nobility of both the more developed provinces and the most elite positions in the central administration points to the primary interests of subsequent chapters. What were the implications of this combination of characteristics during the revolutionary year of 1917? Certainly the combination of elite social status, substantial landholdings, and high administrative office in the most developed provinces of the old Empire must have been disastrous at least for the entire cadre of senior officials if not for the structure of administration itself.

As the next chapter illustrates, evidence of substantial structural change was clear throughout the state administration by the end of 1918. Nowhere was change sharper than in the countryside, where the provincial apparatus was rapidly replaced by peasants' soviets. Because of the close tie between civil administration and landholding, it seems clear that the hottest political issue of the revolution—the fierce demand for land—could not be resolved without destruction of the provincial administration, a move that not only opened the way for but required the establishment of soviets or comparable bodies and left unresolved for a decade the question whether the city or the countryside would win the revolution.

By the same token, the comprehensive evacuation of all high-level positions—provincial and central—can be explained almost entirely on the grounds of social status alone. Whether this explanation by itself is the historically sufficient one is open to question. Succeeding chapters show that there were supplementary reasons for the disappearance of generalist roles. The fact remains, nevertheless, that the disappearance of social elites meant the disappearance of official elites generally. This circumstance opened the way for previously subordinate officials, many of whom were specialists, who were able to demonstrate a "pure"—that is, non-noble, landless—background.

Concluding Observations

It may be helpful to emphasize some connections between the major issues presented in Chapter 1 and the findings of this chapter. First, we have seen that the role of landholding among both peasants and, especially, nobility consistently endured as a pivotal factor in structuring bureaucratic organization and participation in bureaucratic offices. On the one hand, the nobility's weakening grip on the land explains their growing numbers in the civil service in the second half of the century. On the other hand, the wealth and connections conferred by land continued to be critical to the rise of the most highly successful civil servants; the relative importance of large landholdings in accounting for the distribution of gubernatorial offices in the last prerevolutionary generation of provincial administrators underscores the importance of this factor even more sharply. Control of the land would continue to be highly significant in the postrevolutionary era as a determinant both of the social structure of bureaucracy and of its organizational behavior, but the precise nature of its influence would change with the changing nature of landholding itself.

Second, the patterns of bureaucratic development and participation in the bureaucracy reveal at least one aspect of the augmentation of state power in the nineteenth century. The evident objectives of the state—limited land and social reform, urbanization, industrialization, and the professionalization of state service—were chronically at odds with the interests of the landowning nobility. But while the state got more or less what it wanted, this did not absolutely preclude the nobility from advancing some of its own corporate or class interests. Later, after the Revolution of 1917, compromise between the interests of the state and those of specific social groups would continue to be critical to Soviet state-building. Throughout, the compromise of social and political interests, and the integration of policy and personnel that such a compromise permits, allowed the state to make use of the bureaucracy as its major tool for both state-building and development more broadly.

Third, the expansion of a landless, socially inferior, relatively well-educated officialdom appears likely to have made an important contribution to using bureaucracy as a tool for development. The most reasonable explanation for the success of both the emancipation program and the associated policy of land reform is the professionalization of much of the civil service by midcentury. Had a much higher proportion of the civil service consisted of landowners and serfowners, it seems doubtful at best that the emancipation of serfs and post-emancipation administrative development would have proceeded smoothly. Given the Russian proclivity for economic development at the initiative of the state, moreover, much of the industrial-urban development of the later nineteenth century would have been impossible without the landless sub-elite who more and more acquired specialist credentials.

Fourth, the symbiosis between socially inferior officials and the increasing numbers of technical specialists in the central administration points toward a further stage in both bureaucratic and technocratic development at the time of the Revolution of 1917. The fact that specialists were often socially inferior and the evident leverage that specialization gave to career development for the landless contrast sharply with the tendency of generalists to hold controlling positions in both the central and provincial elite administrations and, especially, with the tendency of the landed social elite to hold choice positions. When the rage of the land-hungry population vented itself on the Old Regime after 1917, an inevitable consequence of the fury would be the destruction of the nexus between landholding and elite officeholding. The firestorm of rural revolution consumed elite offices, to be sure, but it seems likely that the prior objective was the destruction of elite officeholders. With one blow, as it were, they were despoiled first of land and then, necessarily, of office and social standing. The events of the rural upheaval, as much as any set of policies introduced by the political leadership, seem to have opened up opportunities for advancement to the socially inferior specialists—if not because they were specialists, then because they were not of the social elite.

Were these opportunities exploited? If so, what were the results? Let us first look at some of the immediate consequences of the Revolution of 1917 for the bureaucracy and its officials.

CHAPTER 3

The First Structural Transformation, 1917–1918

> Attributing the greatest importance to ensuring the uninterrupted activity of all governmental and social institutions, with the aim of maintaining order within the country, and for the successful defense of the state, the Provisional Government has determined upon the necessity of temporarily removing governors and vice-governors from their offices. Administration of provinces will be temporarily discharged for us by the office of a provincial commissar of the Provisional Government, with all the rights currently appertaining in law to the office of governor, while reserving to the presidents of provincial zemstvo authorities the direction of the work of the provincial zemstvo authority. Presidents of district zemstvo authorities will assume the responsibility of district commisar of the Provisional Government while continuing the direction of work of the district zemstvo authority. Departments of police shall be restructured into militias, which [task] must be undertaken by local bodies of self-government.
> Vestnik vremennogo pravitel'stva, Tuesday, 7 March 1917

This chapter's epigraph is taken from a decree on provincial administration issued by the temporary ("Provisional") government. Like many government actions at this time, the decree confirmed a process that was already under way: the wholesale displacement of unpopular officials with representatives of "the people," whatever that might mean in specific cases.[1]

1. N. P. Eroshkin, *Istoriia gosudarstvennykh uchrezhdenii dorevoliutsionnoi Rossii*, 3d rev. ed. (Moscow, 1983), 327–28; A. M. Andreev, *Mestnye sovety i organy burzhuaznoi vlasti (1917 g.)* (Moscow, 1983), 24–26.

The decree marked a departure from the long-established prerogative of the tsarist regime to appoint senior provincial officials. At the same time, this and related government actions completed the dissolution of a substantial portion of a bureaucratic framework that had integrated rural administration into the national and central bureaucratic apparatus. It is true, as Eroshkin has noted, that many of the elements of the tsarist administration remained in place,[2] but the new administrative arrangements that emerged quite early in 1917 resulted in the introduction of provincial, regional, and district representative bodies as oversight groups that broke, for the first time in several generations, the central government's monopoly over provincial appointment and advancement.

The first months of 1917 were marked in Russia by an increasing number of peasant actions directed against officials; against local symbols of wealth, power, and status; and, increasingly, against large landowners of all classes. Such actions ultimately indicated the collapse of central control over rural public affairs, thus setting the stage for the developments in rural, or provincial, administration that were to follow. Students of the land "reform," or redistribution, of 1917–18 have often emphasized the fact that equitable distribution merely spread the poverty of rural Russia more broadly.[3] In 1917, 11.5 percent of Russian peasant households sowed no land, presumably because they had none to sow; by 1919 that figure had declined to 6.56 percent. At the opposite extreme, in 1917 about 5 percent of peasant households sowed more than 10 *desiatins*, while in 1919 that figure dropped to 1.6 percent.[4] The average area sowed by all households was cut nearly in half between 1917 and 1920—from 4.3 *desiatins* to a beggarly 2.2—and as Shanin notes, in some regions average additions to peasant holdings were quite small, less than a *desiatin* per capita; from this he infers widespread peasant disappointment.[5] He may be correct, but the fact remains that because of the despoiling of enclosed farmsteads, more peasants had land, animals, and machinery than ever before in Russian history. Given that these reforms finally resulted in the passing of an unprecedented amount of agrarian wealth into peasant hands, these measures were in many ways among the most important of the entire revolutionary year.

This chapter describes the structural transformation of Russian civil

2. Eroshkin, *Istoriia gosudarstvennykh uchrezhdenii*, 328.
3. Teodor Shanin, *The Awkward Class: Political Sociology of Peasantry in a Developing Society, Russia 1910–1925* (Oxford, 1972), pp. 153–54; John L. H. Keep, *The Russian Revolution: A Study in Mass Mobilization* (New York, 1976), 413–14.
4. Shanin, *Awkward Class*, 53–54.
5. Ibid., 154.

The First Structural Transformation

administration during 1917 and 1918, beginning with an overview of administrative change, passing to a consideration of the many complex changes that occurred in the provinces, and then examining changes at the center. It considers the transformation of offices and legal authority only—the de facto and de jure structure of civil administration; the problem of personnel changes is considered in the next chapter. That the latter changes were complex and important is obvious; administration had fallen into quite different, sometimes alien, hands. That the structural transformation of provincial, regional, and local administration was highly important is less self-evident. Thus, in addition to their descriptive tasks, these two chapters endeavor to show why the changes in regional and local administration were so critical both to the future of the revolution and to the future of Soviet administration.

Administrative Change: An Overview

There are three major components in the story of administrative change in Russia at the time of the Revolution of 1917. First, there are the institutions of the Old Regime, whose survival or demise at both the center and in the provinces must be noted. Then there are newly emergent institutions, more or less sanctioned by the Provisional Government; these include the *zemstvos* and the regional and district land committees, at least in their early stages of organization. Finally, there are such spontaneous or popular organizations as the soviets, the factory committees, and independent peasant land committees. A detailed survey of each element would be long and tedious, especially since their political environment was changing rapidly during 1917. But because the administrative apparatus that began to emerge in 1918 was an amalgam of all of these elements, I shall at least describe their broad development and most important characteristics, both in the provinces and the center, both before and after the Bolshevik seizure of power in October 1917.

The Provisional Government came to power under an informal coalition of members of the national legislature (the Duma) and representatives from people's councils (soviets) in February 1917. The Provisional Government's intention appears to have been to extend to the countryside the political changes instituted in the capital (now Petrograd) and other industrial-commercial centers. That is, its actions and announced policies indicate that the Provisional Government aimed to end autocratic politics and its preoccupation with the Imperial office and person, its anachronistic fixation on class and estate privilege, and its uncritical use of bureaucratic power. Autocratic authority would be replaced by a democratized politics, by the broad use of direct, secret ballots, and by institutions that would be responsible to at least some of the public.⁶ For the provinces this revolutionary agenda would mean such things as election of the governor and other principal officials, some public oversight of police activities, and active participation in politics and administration by organs such as the land assemblies (*zemstvos*) at three of the four recognized administrative levels of government: national, provincial (*guberniia*), and county (*uezd*).⁷

The fourth, or local (*mestnoe*), level of politics and the district assemblies (*volosts*) were problematic because demographically and often economically they tended to be dominated by peasants more than did government at other levels. In general, they showed signs of responding to an agenda different from that of the county and provincial governments. Nevertheless, what had first been a spontaneous move in the countryside to create local administrative authorities was given "a vigorous boost," in the phrase of Dorothy Atkinson, by the Provisional Government's decision on March 19 to legalize district assemblies.⁸ This action, at least from the point of view of the Provisional Government, was complemented in May 1917 by decrees that allowed for the creation of *volost*-level *zemstvo* organizations, aimed at bringing a measure of representative politics to the grassroots of the provinces.⁹ It seems to have been the objective of the Provisional Government that the *zemstvos*, which guaranteed the participation of nonpeasants as well as peasants in local government, would ultimately displace the *volost* assemblies, which were increasingly dominated by peasants only.¹⁰

The Structure of Provincial Administration in 1917

The range of interests evidently supported by the Provisional Government's legislative and constitutional intentions contrasted sharply with organizational and administrative realities at the local level in much of rural Russia. To the peasants, if one judges only from their actions in 1917, the revolution presented very different opportunities

---

6. Graeme J. Gill, "The Failure of Rural Policy in Russia, February–October, 1917," *Slavic Review* 37, no. 2 (1978), 243–44; *Vestnik vremennogo pravitel'stva*, 23 March 1917.

7. Eroshkin, *Istoriia gosudarstvennykh uchrezhdenii*, 331; G. A. Gerasimenko, *Nizovye krest'ianskie organizatsii v 1917—pervoi polovine 1918 godov: Na materialakh Nizhnego Povolzh'ia* (Saratov, 1974), 46–47. See also *Vestnik vremennogo pravitel'stva*, 21 and 25 May 1917.

8. Dorothy Atkinson, *The End of the Russian Land Commune, 1905–1930* (Stanford, Calif., 1983), 150.

9. *Vestnik vremennogo pravitel'stva*, 25 May 1917.

10. Atkinson, *Russian Land Commune*, 150.

from those seen by city liberals associated with the Provisional Government or the zemstvos.

It is not exactly that peasants were uninterested in the liberal institutions so dear to the educated and well-to-do politicians of the Provisional Government; peasants were apparently interested in guaranteed freedom of assembly and speech, for example. The first would allow them to meet, whatever the legislative intent of Petrograd, for purposes of legal collective decision-making at the village and district levels of government, close to their homes and their traditions and far away from the oversight of provincial, landowning politicians and bureaucrats. The importance of this right may be more clearly understood if we remember that the one elected organization with legal authority, the zemstvo, had previously met only at intermediate—county and province—administrative levels, not at the micropolitical level where peasants were strongest or at the macropolitical level, in St. Petersburg, where the power of the landlords might compete too effectively with that of the bureaucrats. Local administration in these circumstances had tended to be dominated by appointed officials—the police; the conservative, landed, and powerful land captains; and judicial authorities—or by a tame village assembly, the skhod. Freedom of assembly, before 1917 had run its course, became an important peasant political right and a foundation stone for a peasant-influenced local and district administration.

Similarly, the importance to the peasantry of freedom of speech and of the press during the first revolutionary year was evident in the emergence of radical provincial newspapers, the rise of political splinter groups, and peasant voting behavior. This situation would change later on, but in 1917 peasant assemblies heavily supported political parties that had little or no national standing.[11]

It seems clear, however, that peasants were even more interested in other things. Or, to make a logical inference from their behavior in March–September 1917, they were interested in liberal political guarantees only as a means to attaining what were in their view more important ends. Thus, whatever the central government did, peasants took two steps—often on their own, sometimes at the urging of "outside" agitators—that were critically important for future administration in the provinces.[12]

First, as the weeks passed in 1917, peasants transformed their *volost*

11. Launcelot A. Owen, *The Russian Peasant Movement, 1906–1917* (London, 1937), 155–59; Gerasimenko, *Nizovye krest'ianskie organizatsii*, 224–27.
12. Gerasimenko, *Nizovye krest'ianskie organizatsii*, 231ff; A. D. Maliavskii, *Krest'ianskoe dvizhenie v Rossii v 1917 g. mart-oktiabr'* (Moscow, 1981), 249–63, App. Table 2.

and village assemblies into organizations that included only peasants and that acted in the peasants' interests, often on an exclusionary and prejudicial basis.[13] That is, these organizations frequently and quickly became the basis for pseudolegal peasant action in areas of great political controversy, such as questions of who should control land, agricultural labor, and agricultural products. These organizations often excluded not only the great landowners, nobles, and merchants but even peasants who held enclosed and individually owned land (*khutorshchiki* and *otrubshchiki*).[14]

Second, peasants began almost immediately to take actions (*dvizhenie*), village by village, against independent landowners and their property. These measures preempted legal and administrative authority previously reserved to other levels of government and were often ratified and administered by the *volost* or the village assemblies. Usually the targeted landowners were nobles, but increasingly, as the year 1917 wore on, they included all independent or non-communal landowners, whether peasant, bourgeois, or noble.[15]

Table 10 is based on data presented by Gill, who relied on Kotel'nikov and Meller.[16] As the table indicates, in the earliest months of the revolution, peasant action consisted mainly of confrontations between persons, highlighting antagonism over private possessions and, perhaps, social status (in the "physical violence" and "other" categories). As agricultural activity picked up, however, so did the focus of peasant action on, first, land seizure and then crop seizure. This shift lends credence to the argument that the activity was keyed to the natural rhythm of the agricultural season.[17] By July-August these two categories combined accounted for more than half of peasant actions and were apparently coordinated with the harvest.[18] After harvest, land and crop seizures subsided in favor of such activities as illegal timbering, unilateral lowering of land rents, and so forth. It should be noted, however, that in the view of some commentators this was a respite before the great storm of land seizure that occurred in the winter of 1917–18.[19]

13. Gerasimenko, *Nizovye krest'ianskie organizatsii*, 56, Table 2.
14. Ibid., 53ff; Atkinson, *Russian Land Commune*, 150–51.
15. Maliavskii, *Krest'ianskoe dvizhenie*, 178–84, 309–27; Shanin, *Awkward Class*, 150–51.
16. Graeme J. Gill, "The Mainsprings of Peasant Action in 1917," *Soviet Studies* 30, no. 1 (1978), 68–70; K. G. Kotel'nikov and V. L. Meller, *Krest'ianskoe dvizhenie v 1917 godu* (Moscow, 1927), Table 1.
17. Gill, "Mainsprings," 76; Owen, *Peasant Movement*, 139.
18. Owen, *Peasant Movement*, 139; Atkinson, *Russian Land Commune*, 162, Table 12. Atkinson says that Soviet scholars tend to attribute more importance to politics than to nature in explaining changes in both patterns and, especially, intensity of peasant activity (163). But see my comments further on.
19. Keep, *Russian Revolution*, 394–95.

TABLE 10.
Patterns of peasant action, 1917 (specific events as % of the total)

	March	April	May	June	July	August	September	October
Land seizures	2.6%	24.9%	34.3%	37.0%	34.5%	35.8%	23.6%	18.2%
Crop seizures	7.7	1.9	2.6	7.8	23.7	22.2	11.9	11.3
Physical violence	59.0	20.7	17.3	12.7	11.4	21.2	32.0	30.9
Timber seizure	25.6	20.2	19.9	17.9	10.9	11.0	26.7	32.6
Other	5.1	32.4	25.8	24.6	19.5	9.9	5.7	6.4

Source: Calculated from Graeme J. Gill, "The Mainsprings of Peasant Action in 1917," Soviet Studies 30, no. 1 (1978), 68 n.12.

By year's end, peasant actions throughout the main agricultural regions of the country numbered in the thousands. Maliavskii, in an effort to identify and quantify peasant actions of all kinds from the beginning of March 1917 to the end of October, arrived at a total of 16,298, including everything from the murder of landlords to land seizures.[20]

Even if actual incidents of physical violence were relatively rare, the constant threat of it was no doubt terrifying to landowners and their families.[21] But the actions having the longest-lived consequences and bearing the most profound implications for administrative change were, clearly, land seizures coupled with the creation of the peasant committees that ratified these seizures. By the end of the season, such actions had been so pervasive and effective that considerable leveling had already been accomplished—to some degree throughout the country and very intensely in certain provinces—although more would soon be achieved. It is important to recognize that land reform, even before the Bolshevik seizure of power and the famous Land Decree, was linked in peasant experience to the creation and subsequent activity of local bodies, the village *skhody* and the *volost* assemblies.

When the Bolsheviks seized power from the Provisional Government in Petrograd in October 1917, one of their very first acts was to sponsor and publish the Land Decree, which gave their support and that of their political allies to immediate seizure by the peasants of large private and state landholdings. It is commonly proposed that Bolshevik support for land expropriation helped to forge a community of interests between peasants and urban workers.[22] The argument has been forcefully advanced, however, that peasant actions had little to do with the urban revolutionary politics of 1917–18 and rather more to do with peasant customs, especially those arising out of a communal tenure system that attempted to adjust landholdings in proportion to the number of workers or "eaters" in a family—in other words, that most of the Russian peasants would, in any case, have wanted to redistribute land to achieve some kind of egalitarian tenure system and to place its management under the control neither of individuals nor even of families but of the repartitioning commune.[23] Let us explore further the implications of this interpretation.

In one sense, the questions of how much land peasants obtained and

20. Maliavskii, Krest'ianskoe dvizhenie, App. Table 1; see also Kotel'nikov and Meller, Krest'ianskoe dvizhenie v 1917 godu, Table 1.
21. Keep, Russian Revolution, chaps. 15, 16; Owen, Peasant Movement, chap. 4. I make my judgment that such incidents were rare relative to what might have been and relative to the destructiveness of other modern "peasant wars"; see Eric Wolf, Peasant Wars of the Twentieth Century (New York, 1969).
22. Owen, Peasant Movement, p. 143; Maliavskii, Krest'ianskoe dvizhenie, chap. 3.
23. Gill, "Mainsprings," 64ff.; Shanin, Awkward Class, 151.

perhaps even who was despoiled of land are moot. The Land Survey of 1916 makes it clear that before 1917 large numbers of peasant farm families subsisted entirely without land, whereas substantial amounts were accounted for by a small but significant percentage of farms that controlled sown areas in excess of 100, 500, or even 1,000 *desiatins*.[24] Whether this was a sign of systematic exploitation of poor peasants or of general rural Russian poverty is a question I am not prepared to go into here. It should be noted, however, that ten years later, according to the findings of the Central Statistical Administration, essentially all land had become peasant communal land or state land.[25] In the post-1917 period, moreover, the general trend in landholding throughout the country was for families or farms with little or no land to decline as a proportion of the total, while those with relatively large holdings—measured on the order of 25 *desiatins* or more—declined precipitously in the early postrevolutionary period and then either stayed low or rose only slightly.[26] The implication of this trend is that all large landholders—including nobles and peasant holders of enclosed land—disappeared from the stage in the course of revolutionary land reform.

The kinds of activities in which peasants engaged, where and in what political context, are perhaps even more important than land redistribution to any attempt to understand administrative development, or at least more relevant to my central concern of how local administration was affected by the revolution. In general, with the important exception of the left-bank Ukraine, regions with relatively high rural populations and large areas of comparatively productive, or valuable, agricultural land were most prone to peasant actions of all kinds in the early stages of the revolution. One can press this point further. Table 11 shows total "events" and total events per capita of rural population in March–October 1917. In two groups of provinces—the Central Black Earth and Volga groups—we find substantially higher ranges of activity both absolutely and per capita. There were, for example, more than four times as many peasant actions per capita—that is, cases of peasants taking the initiative against owners of enclosed land—in the mid-Volga provinces of Penza, Simbirsk, Saratov, and Kazan as in the agriculturally less productive Byelorussian region including Minsk, Mogilev, Vitebsk,

---

24. S. M. Dubrovskii, *Sel'skoe khoziaistvo i krest'ianstvo Rossii v period imperializma* (Moscow, 1975), chap. 7.
25. Central Statistical Administration USSR, *Itogi desiatiletiia Sovetskoi vlasti v tsifrakh, 1917–1927 gg.* (Moscow, 1927), Otd. V, *Sel'skoe khoziaistvo*.
26. Ibid., Table 4, "Raspredelenie krest'ianskikh khoziaistv po rabochemu skotu, korovam, i posevnym ploshchadiam v protsentakh k obshchemu chislu khoziaistv" (Distribution of peasant households, in percent, by draft animals, cows, and arable land). Two regions in which this pattern did not hold were the Northern Caucasus and Crimea, where land allotments in excess of 16 *desiatins* actually rose during the mid-1920s.

TABLE 11.
Aggregate and per capita incidents of peasant action, March–October 1917

Region	Total incidents	Incidents per capita
Northern	15	.006
Urals	426	.057
Baltic	85	.042
Belorussia	785	.092
Lithuania	46	.009
Black Earth	4,440	.272
Volga	3,614	.403
Southwest	1,736	.151
Ukraine (left bank)	750	.083
South Steppe	725	.054
Southeast	926	.097

Source: A. D. Maliavskii, Krest'ianskoe dvizhenie v Rossii v 1917 g. Mart-Oktiabr' (Moscow, 1981), app. Table 1.

and Smolensk. In these same two groups of provinces, data from A. G. Rashin and other sources show both comparatively high concentrations of rural inhabitants and high land values.[27]

Obviously, these findings do not argue that variation in soil quality was the cause of peasant unrest in 1917. They do call into question the idea that peasants were only responding to the "natural rhythm" of the harvest season, and they confirm the expectation that as the value of land rose, so did peasant demand for it.

The evidence suggests, more pointedly, that peasant activism in 1917 was linked not just to seasonal factors, as suggested by Gill and Owen, but to economic—if not political—factors, as suggested by some Soviet writers. In my view, therefore, the economic and demographic motivation of peasant actions does seem to emphasize the independence of revolutionary actions in rural areas from those in Petrograd or other large cities. It was not merely a matter of tradition—though tradition was certainly important; it was not simply a matter of nature and natural rhythms—though it was already clear that food was going to be in short supply by the end of the harvest season. It was also a matter of raw economic and political self-interest as seen from the viewpoint of the peasants.

If this was the case, what were the implications for the creation of new political and administrative organizations? The answer seems to

---

27. For urban-rural population data, see A. G. Rashin, Naselenie Rossii za 100 let (1811–1913 gg.): Statisticheskie ocherki (Moscow, 1956), Tables 22, 57; for approximate land values and loan amounts, see N. A. Proskuriakova and A. P. Korelin, Materialy po istorii agrarnykh otnoshenii v Rossii v kontse XIX—nachale XX vv: Statistika dolgosrochnogo kredita v Rossii (Moscow, 1980), 3–29, Table 1.

be that independent peasant actions and the communal and *volost* organizations needed one another for purposes of legitimacy. These local organizations and the groups of peasant leaders who ran them gave a patina of justice and right to land and movable property seizures; by the same token, independent peasant actions provided the substance for local organizations' agendas. The evident inference, then, is that the continued existence and operation of these organizations were essential to the survival of "the Revolution" in the countryside.

Any attempt to return to the *status quo ante* administrative system would entail restricting or abolishing local peasant organizations, and peasants would inevitably see such an act as the first step in rolling back land reform. So just as Russian prerevolutionary institutions were closely linked to prerevolutionary land tenure, so postrevolutionary, peasant-dominated institutions were linked to land reform and the new order of land tenure.

Given the foregoing reasoning, the idea that the peasant revolution was an appendage of the urban revolutionary experience—a favorite interpretation of Soviet historians—is difficult to accept.[28] By the same token, the concept put forward by Gill and Owen—that peasants were following their own values—seems helpful. It was the commune that seized land, and it was the village or *volost* assembly—often the same people who belonged to the commune—that legitimated the seizure. Thus, as the revolution matured, so did the microlevel organizations.

As the months passed, these local organizations assumed administrative roles. Gerasimenko notes the creation by the assemblies of administrative executive committees and describes their functions in the Lower Volga region: law and order, social services, fiscal management, and the like.[29] At the same time, there is evidence that these assemblies crowded out not only the *zemstvos* but even local rural soviets. By the autumn of 1917 only 2 percent of the *volosts* in the Lower Volga provinces of Saratov and Astrakhan had peasant soviets; in the country as a whole the figure was only 1.3 percent.[30] Clearly, the favored peasant revolutionary forums were the communal and *volost* assemblies, and the favored means of handling local administrative problems were the executive committees of the *volost* assemblies.

With the passing of 1917 and the winter of 1918, the Bolsheviks made a concerted effort to extend the soviets into areas under the military

28. Maliavskii, Krest'ianskoe dvizhenie, chap. 3; E. G. Gimpel'son, Rabochii klass v upravlenii Sovetskim gosudarstvom: Noiabr' 1917–1920 gg. (Moscow, 1982), 182–201. Many of Gimpel'son's data come from 1918 and later, but I read him to mean that worker leadership was a fixture of the entire revolutionary period.
29. Gerasimenko, Nizovye krest'ianskie organizatsii 254, Table 19.
30. Ibid., 73–74, Table 6.

control of the revolutionary government.³¹ This drive took the form of (1) assuring the creation of soviets and their executive committees in all provincial capitals; (2) amalgamating worker, soldier, and peasant soviets at the provincial and county level wherever they were separate (usually peasant soviets were separate); and (3) creating soviets in districts and villages where they were infrequently found. These steps were taken through the intervention of agitators: Bolsheviks, members of soviets from more urbanized areas, veterans of the army. What they achieved, broadly, was the fabrication of a skeleton of provincial administration that was sufficiently soviet and sufficiently radical to give the Bolsheviks ongoing administrative control. At the district and local levels in the postrevolutionary era, however, even though the system would have a soviet structure, it would be numerically dominated by peasants throughout the 1920s.

For present purposes, we should merely note that by mid–1918 the rough framework of what was to become Soviet provincial administration was in place. In the central agricultural, industrial, and Lower Volga regions this framework consisted of organizations composed of elected representatives at the village, district (volost), county (uezd), and provincial levels. As time passed, where these organizations had been either zemstvos or peasant-dominated land committees, they increasingly became soviets founded under the prod of agitators from other soviets, the Bolsheviks, or their political allies.

This structure was gradually brought under the control of leadership groups favorable to the interests of the Bolshevik government in much the way that soviets in Moscow and Petrograd had fallen under Bolshevik control several months before.³² Working in conjunction with the soviets were executive committees (ispolnitel'nye komitety) that assumed the responsibility of day-to-day management of public affairs: tax collecting, maintenance of order, marriage licensing, and so forth. Because of the special circumstances of the worsening military situation and of famine, these executive committees were supplemented by organizations that concentrated on special problems: labor supply and management, food collection and distribution, and land distribution. This network of soviets, executive committees, and special committees was, as time passed, increasingly integrated into a national soviet ad-

31. Keep, Russian Revolution, chap. 33; Gerasimenko, Nizovye krest'ianskie organizatsii, 231–32.
32. Diane Koenker, among others, insists that this process was subtle, complex, and uncertain; see Moscow Workers and the 1917 Revolution (Princeton, N.J., 1981), 268, chap. 7. T. H. Rigby, Lenin's Government (Cambridge, 1979), 170, refers to the "drastic decline" in the influence of the soviets and the tendency of the Moscow leadership to rely on the Cheka (political police) and other powerful agencies as means of asserting central control.

ministrative apparatus, dominated by the Bolshevik Party membership and, especially, by the commissariats and special administrations that were being established in the new capital, Moscow.

An administrative pattern in which initiative came from the center and was bureaucratically executed in the provinces had long been established in Russia. The broad national organizational infrastructure, political ambition, and social expectations that would have sustained some alternative approach—such as a federalist system—did not exist. Millions of everyday events in the lives of provincial Russians, whether urban or rural, were simply not consummated or considered ratified unless they were approved by an official representing the state, the central state apparatus. To imagine that this circumstance could change in a matter of months is to give too little weight to the role of authority in human social life.

At the local level, however, the soviet apparatus was obliged to exist side by side, throughout the 1920s, with peasant communes (*skhody*). These organizations, also blessed by long-established tradition and often the source of emerging peasant radicalism, had been given a new lease on life now that land reform was a reality. Structurally, then, at the local level—the point at which government made contact with millions of peasants—Soviet power was at best shared with a deeply rooted organization, the commune. And both soviet and commune, we are tempted to think, were obliged to draw on the same personnel resources for their administrators and representatives.

Central Structural Change: Petrograd and Moscow

In the aftermath of revolution, debate among the victors is inevitable. No amount of prerevolutionary planning or plotting can take account of all postrevolutionary contingencies. Nor does it ever seem to be the case that the world wrought by revolution corresponds with prerevolutionary expectations.

In Soviet Russia during the 1920s, debate and competition for control over policy formulation and implementation were the order of the day.[33] In the revolutionary era, however, even if opinion was not unanimous, debate was often irrelevant simply because action was imperative. An army had to be created and fielded to fight the civil war and to stave off Polish and interventionist forces. Foreign relations had to be maintained or reestablished. Domestic administration had to continue. Consequently, the postrevolutionary "solutions" to problems of policy and

---

33. Moshe Lewin, *Political Undercurrents in Soviet Economic Debates: From Bukharin to the Modern Reformers* (Princeton, N.J., 1974), chaps. 2, 4.

administration in many areas were concocted on the run—amid debate, to be sure, but without extensive forethought. Perhaps partly for this reason the administrative systems that emerged after the Revolution of 1917 bore the unmistakable imprint of compromise. They were clearly the outcome of a mélange of revolutionary objectives, the accumulating demands of the present, and the irresistible weight of the past.

Urgency in the formation of new government institutions or revision of old ones is illustrated by the fact that by the end of 1917, administrative structures were renamed, redefined and restaffed by the Provisional Government (as shown by Likhachev and Eroshkin as well as by the official press of the period).[34] Nevertheless, social, economic, and political problems continued to accumulate, and solutions were badly needed. The result was a compromise structure, with modified objectives, that used both some of the personnel and some of the administrative structures of the past.

In his study of the early Soviet central administration and government, T. H. Rigby remarks on the striking similarities between the government created after October 1917 and the one in place before either the February or the October Revolution. As it was, Rigby wrote, "the structural changes were scarcely greater than those sometimes accompanying changes of government in Western parliamentary systems. The personnel changes were greater and could perhaps be compared with those occurring in Washington in the heyday of the 'spoils system.' "[35] In broad terms Rigby may be correct. If one thinks of the changes that *might* have been attempted, even in the short run, those that *were* made seem relatively mild, lending credence to Lenin's statement that "our state apparatus, with the exception of the People's Commissariat for Foreign Affairs, represents in the highest degree a hangover of the old one, subjected to only the slightest extent to any serious change."[36]

From another perspective, however, beginning quite early in the revolutionary era, the changes seem rather more profound. For one thing, the revolutionary leadership of both the Provisional Government and

---

34. M. T. Likhachev, "Gosudarstvennye glavnye i osobye komitety vremennogo pravitel'stva," *Voprosy Istorii* 2, (1979), 30–41; Eroshikin, *Istoriia gosudarstvennykh uchrezhdenii*, 308–44 and Table, "Izmeneniia v sostave Vremennogo pravitel'stva," 325–26. Some senior personnel changes are recorded in *Vestnik Vremennogo Pravitel'stva*. For an interesting comparative perspective, see Rigby, *Lenin's Government*, chap. 15.

35. Rigby, *Lenin's Government*, 51. Rigby notes his indebtedness, in making these comparisons, to M. P. Iroshnikov's *Sozdanie Sovetskogo tsentral'nogo gosudarstvennogo apparata: Sovet Narodnykh Komissarov i Narodnye Komissariaty, Oktiabr' 1917–Ianvar' 1918 g.* (Leningrad, 1967).

36. V. I. Lenin, *PSS*, 5th ed. (Moscow, 1958–65), 45:383, quoted in Rigby, *Lenin's Government*, 51.

The First Structural Transformation

the new Soviet government laid the foundations for several new administrative structures in the months following the abdication of the tsar. For example, the Department of Police, centrally controlled by the Ministry of Internal Affairs, was abolished early in the career of the Provisional Government and replaced by a "militia" designed to be locally controlled. New ministries—later commissariats—such as those of Labor, Food Supply and Social Welfare, were established. This process of administrative transformation continued after the October Revolution.

Perhaps it goes without saying that even under the Provisional Government (March to October 1917) personnel changes were extensive. Eroshkin lists forty-eight replacements in seventeen ministerial-level offices between March 2 and October 25.[37] But more on this important matter later. It was the *structure* of both the central and provincial governments, under the impact of changes introduced by the Provisional Government and by the Bolsheviks, that most obviously acquired a new look. The total number of institutions of central administration increased—as it would continually do in the Soviet era—with the ministries founded by Alexander I being subdivided into a larger number of new, more specialized organizations.

To take but one example, the pre-1917 Ministry of Internal Affairs was, by the end of 1918, replaced by five commissariats: Posts and Telegraphs, Social Welfare, a more narrowly structured Internal Affairs (ultimately regaining control over the militia, or civil police, that had been separated from it by the Provisional Government), Nationalities, and Health Preservation. To these must be added an independent Statistical Administration, the political police (V. Ch. K., or Cheka), and two commissariats—Foreign and Internal Trade, and Agriculture—which fell heir to portions of the old Ministry of Internal Affairs and of other ministries. There were probably more; this list, moreover, does not take into account the intermediate steps through which the formation of these commissariats sometimes passed. For example, the Provisional Government created new organizations, such as the Ministry of Welfare already noted above, which were transitional from the tsarist structures to the Soviet commissariats.

Although it is more instructive and important for us to understand the kinds of changes that occurred in the personnel of these agencies, it is also helpful to recall the overall pattern of change that ministerial government underwent from 1801 to the mid-twentieth century. As we have already seen, specialist roles began to develop well back in the nineteenth century. But specialist ministries (or commissariats)—agen-

37. Eroshkin, *Istoriia gosudarstvennykh uchrezhdenii*, 325-26.

cies in which budgets, personnel appointments, and most day-to-day policy decisions are under the control of professional specialists—are a twentieth-century phenomenon.

Obviously, it is not coincidental that the growth of independent, professionally and technically specialized administrations began with the first generation of Russian industrialization at the end of the nineteenth century (the Ministry of Agriculture and the Ministry of Trade, for example), and that this growth became truly explosive with the second industrial spurt beginning at the end of the 1920s. As Alfred Rieber shows, the satisfaction of social, political, and economic demands imposed by urbanization and industrialization was not achieved by private entrepreneurship or independent local initiative.[38] The usual presumption in nineteenth-century Russia was that the state had the right of first refusal whenever there was a question of managing new or extended power, whether the power was classically political, economic, or merely social. This meant that finding solutions to the problems created by economic development or the lack of development was the primary responsibility of formal governmental structures. So in spite of all the apparent change that is associated with the revolution, it should not surprise us that when Russia once again focused on economic questions and matters affecting the socioeconomic order, the natural tendency was to look to the state for organizational solutions.

It is not a matter for concerted attention here, but I should note that the tradition of presumptive central state authority in Russia explains, in broad terms, the ascendance of centralist institutions such as the Council of People's Commissars (SOVNARKOM) and the Central Committee of the Bolshevik/Communist Party. As they developed, these institutions concentrated administrative power at the center over a geographically extended hierarchical apparatus. This structure was in the best Russian political tradition. The All-Russian Central Executive Committee of Soviets, by contrast, lost the struggle for power with SOVNARKOM precisely because of the diffused nature of its writ of authority. It was an organization that ultimately required, of all things, popular support to substantiate its claim to supreme control. In Russia such a claim was simply too alien and suspect to be taken seriously by very many people for more than a short time.[39]

At the same time, however, the state's right of first refusal in all

---

38. Alfred J. Reiber, *Merchants and Entrepreneurs in Imperial Russia* (Chapel Hill, N.C., 1982).

39. For discussions of the competition between SOVNARKOM and the Central Executive Committee, see Rigby, *Lenin's Government*, 164–90; and J. L. H. Keep, ed., *The Debate on Soviet Power: Minutes of the All-Russian Central Executive Committee of Soviets, Second Session* (Oxford, 1979).

matters relating to power and authority should not be taken to mean that social and economic experiences had no bearing on governmental changes. Quite the contrary: the causal force often flowed in the opposite direction. Structural change, like change in the complement of personnel at all levels, was the product of decisions by political leadership but also of demands imposed from below by the social and economic experience of millions of people.

Finding the New Soviet Structure: An Example

Perhaps we can see this point more clearly by examining in some detail the 1917 and 1918 transformation of administrative functions in health administration. This case is interesting because health policy was (and is) relatively neutral from a political standpoint, a kind of residual issue—important if there is a specific crisis but otherwise the kind of thing political leadership gets around to "later." The major point of contention, the socialzation of health or its transformation into a government-controlled resource, was a foregone conclusion under the Bolsheviks. But the way in which the People's Commissariat for the Preservation of Public Health (NARKOMZDRAV) took shape and the rate at which it emerged into a full-fledged body illustrate many of the characteristics of Soviet administrative development without raising too many of the extremely tendentious issues that attach to the formation of, say, the police, the Red Army, or the foreign affairs administration. Health administration, moreover, does not beg the question of technocratic development in the way that economic and industrial administration do; one need not argue that the Commissariat of Health developed as it did simply because of the leverage of some of its operatives.[40]

In the early twentieth century, in spite of the broad role of the Imperial state apparatus in medical and health care before 1917, the legislation that created the apparatus did not conceive of these functions as a coherent whole. In the nineteenth century the larger administrative objectives of the dominating agency had defined organizational patterns, and these in turn operated under very broad or global adminis-

40. Other views of commissarial formation abound: Iroshnikov, *Sozdanie sovetskogo tsentral'nogo gosudarstvennogo apparata*, pt. 2; Rigby, *Lenin's Government*, chaps. 4, 10–12; Sheila Fitzpatrick, *The Commissariat of Enlightenment: Soviet Organization of Education and the Arts under Lunacharsky* (Cambridge, 1970), chap. 2; M. B. Kerim-Markus, "Kadry NARKOMPROSa v pervyi god Sovetskogo gosudarstva," *Istoricheskie zapiski Akademii Nauk SSSR* 101 (1978), 72–99; E. N. Gorodetskii, *Rozhdenie Sovetskogo gosudarstva, 1917–1918 gg.* (Moscow, 1965); V. Z. Drobizhev, *Glavnyi shtab sotsialisticheskoi promyshlennosti* (Moscow, 1966); V. M. Lesnoi, ed., *Sotsialisticheskaia revoliutsiia i gosudarstvennyi apparat* (Moscow, 1968).

trative goals better thought of as sentiments than as organizational objectives. Thus, as we have seen, legislation defined the objectives of the Ministry of Internal Affairs in terms of attaining the "good order, health, and well-being" of the whole empire, a formulation that could encompass almost any sort of administrative activity. With respect to health agencies, there were no organizations for administering pediatric medicine or sanitation research. There was school medicine (under the Ministry of Education), welfare (under the Ministry of Internal Affairs), and epidemic control (also under the Ministry of Internal Affairs as well as an independent imperial commission). At the provincial and local levels also, medicine was divided into medical inspection, police medicine, and so on.

A major change began in 1917 as most medical and related health care agencies were grouped together. The result was that not only were medical specialists administering their own bureaucracies and reporting to medical professionals at higher administrative levels, but their functions were divided along professional or technical lines. After July 1918 the Soviet political administration developed an organizational framework that gave substantially increased scope to the influence of medical professionals. The new approach structured local and provincial organizations along lines of medical specialization, and these were integrated for the first time into a single organizational unit, a commissariat, under the control of two physicians-become-bureaucrats.

The process whereby the large, complex organizations of the tsarist era became the narrow, operationally specialized agencies of the early postrevolutionary era was not the sort of clean, swift restructuring that an authoritarian interpretation of Russian politics would lead one to imagine. The transformation in public health was a protracted, complex affair involving the conflicting interests of holdovers from the tsarist bureaucracy, "public physicians," Bolshevik ideologues, and an urban public terrified of cholera, typhus, and typhoid. Not least, of course, the transformation also touched the interests and ambitions of posturing politicians.

To make matters more complicated, reorganization of the major administrative entities of the central government actually extended across three different political regimes (tsarist, Provisional, and Soviet) and engaged the interests of many would-be leaders of reform. In health administration the reorganization included programs pushed by a senior official of the Ministry of Internal Affairs Medical Council, G. E. Rein, and instituted with questionable legal authority against opposition. It also included the Provisional Government's program, which ended by abolishing Rein's new agency and enhancing the influence of medical operatives in the *zemstvos*, the so-called public physicians.

The First Structural Transformation

Last, of course, it engaged the attention of a beleaguered Soviet government and its leader; Lenin attempted to apotheosize the struggle for public health with the curious and famous aphorism "Either the louse will defeat socialism or socialism will defeat the louse!"

The impulse to reorganize civil administration, as exemplified by the history of the health administration, was shared among reformist and radical political leaders, including V. I. Lenin himself. It seems unlikely that the Bolsheviks initially envisioned any sort of strong central administrative body, given the formulation published by Lenin in the *April Theses* (1917). As expressed in that document, the role of medical professionals seemed destined to be limited to *providing* medical care rather than administering it, the same distinction that had generally worked to the disadvantage of physicians before the revolution.

There is a substantial literature that may be classified as "Lenin's ideas about health care," and it is natural to assume that one may discover the Bolsheviks' plans for a postrevolutionary health program by reading it. This subdivision of Leniniana is enormous; in the third edition of his book on Lenin and health, B. M. Potulov lists approximately 700 works on the subject.[41] Of these, a considerable number are either by Lenin or explicitly concerned with describing his role in the development of health care.

Nevertheless, after a careful examination of Lenin's published prerevolutionary writings, one must conclude that he gave little forethought to the question of health preservation with respect to caring for the sick, educating medical professionals, or controlling epidemics. The vast majority of his writings and speeches that are said to concern the preservation of health (*zdravookhranenie*) deal with this subject only in the general sense of social welfare and working conditions. In this way his approach is similar to that of the Party project for health care and the preservation of health which the leadership presented to the 1903 Russian Social Democratic Congress. Although some of the ideas and specific objectives of this program did change between 1903 and 1917, that project is nevertheless illustrative of the views of both Lenin and the Party.[42]

The statements of 1903 do not focus on disease control, hospitals, medical schools, and the distribution of physicians. Instead, these pronouncements concentrate on the symbolic welfare issues of the workers' movement at the turn of the century, reflecting Lenin's own point of view. These included the length of the working day, general working

---

41. B. M. Potulov, *V. I. Lenin i okhrana zdorov'ia sovetskogo naroda*, 3d ed. (Moscow, 1969).
42. *KPSS v rezoliutsiiakh i resheniiakh s"ezdov, konferentsii i plenumov TsK* (Moscow, 1954), 1:41ff.

conditions, night work for juveniles, maternity leave, and similar issues. Another major preoccupation is workers' compensation in cases of illness or accident. All of this would not be particularly remarkable but for the fact that less than a year after coming to power, the Bolshevik wing of the Social Democrats with Lenin at their head was at the political center of the first national health ministry in world history, an institution whose fundamental purposes were the prevention of illness and the care and rehabilitation of the sick.

Although it is hard to see, under the Leninist scheme of things, how individual professionals could operate independently of institutions such as factories, hospitals, or the state itself, Lenin's prescriptions for the future state dwell little on these institutions or on individual roles. Jeremy Azrael notes that Lenin, for one, distinguished between "accounting and control over production" and the "management of production."[43] Such a distinction may help the reader to comprehend statements in *The State and Revolution* that otherwise seem either naive or disingenuous. It is in that work that we find Lenin's familiar statement that the "chief things necessary for the smooth and correct functioning of the first phase of communist society" are nothing more than "bookkeeping and control."

This was hardly Lenin's only word on the subject of expert administration in the postrevolutionary era. He wrote of the importance of the roles of technicians and managers and stated that, as Azrael quotes him, "the victorious Bolsheviks would give all those who had experience in organizing banks and enterprises 'their suitable and usual work.'" Thus, if Lenin failed to concern himself in detail with the roles of technically qualified specialists—whether in industry or health care—it was not because he thought there would be no demand for their services. He devoted little space to discussion of the form that controlling and providing for medical care would take not because he believed that such questions would simply disappear after the revolution but because they were less salient than questions of *political control* over the technicians, specialists, managers, and their organizations. As we saw in Chapter 1, such questions are chronically controversial in any consideration of the creation of technocracy. To these questions, Azrael makes clear, Lenin devoted considerable attention, and he was not alone in attaching great importance to political control over bureaucracy.[44] Nevertheless, one is justified in seeing him as the

---

43. Jeremy Azrael, *Managerial Power and Soviet Politics* (Cambridge, Mass., 1966), 13–16, 43–47.
44. Ibid., 23–27.

first of many Soviet political leaders to be disappointed in the potency of that control in the 1920s.

The Eighth Party Congress in 1919 made a formal commitment to specific ambitious public health and medical care objectives, summarizing these objectives as measures to be taken in the interest of workers: (1) sanitizing of populated areas; (2) establishment of social nutrition on scientific-hygienic principles; (3) organization of measures against the spread of illness; (4) establishment of sanitation laws; (5) struggle against social diseases such as tuberculosis, venereal disease, and alcoholism; (6) guarantees for universal, free, and qualified clinical care and medication.[45]

Rarely in the history of Party meetings did public and clinical health problems claim so much time and space. Yet these pronouncements were very much after the fact; by that time the major characteristic features of a centralized health management system were in place and operating. And to make matters more confusing, during the months between the issuance of Lenin's *April Theses* and the convening of the Eighth Party Congress, regional and local health administration actually dissolved into a collection of apparatuses under the control of a bewildering variety of political authorities.

The Reemergence of a Central-Regional-Local System

In the first phase of development of Soviet medical administration the reorganization and control of operational agencies devolved into the hands of regional and local soviets, their executive committees (*ispolnitel'nye komitety* or *ispolkoms*), and their nonmedical appointees. This development, though it falls far short of professional bureaucratic control, was characteristic of the civil war era.[46] The Commissariat of Health was not founded until July 1918, and it asserted its authority over local and regional medical organizations only gradually after that.

If, in considering the interaction of political authorities and administrative development, one gets the impression that the two were not in perfect synchronization, it was not the first instance of a lack of coordination between political authorities and the administrations that were supposed to serve them. In the circumstances of 1918, Lenin and

45. KPSS v rezoliutsiiakh (Moscow, 1960), 2:36–77.
46. Marc Ferro, *Des Soviets au communisme bureaucratique* (Paris, 1980), 110–11, 119; T. H. Rigby, "Early Provincial Cliques and the Rise of Stalin," *Soviet Studies* 33, no. 1 (1981), 13–15.

other Bolshevik politicians shared the view that the new Soviet administration responded poorly to the theoretically powerful Bolshevik political machine.

In the health field, the Soviet government began pressing as early as November 1917 for local soviets to create divisions that would deal specifically with health problems—the medico-sanitary divisions.[47] Articles appealing for their creation appeared in *Izvestiia sovetskoi meditsiny* (News of Soviet medicine) during 1918. As described by this journal, the official voice of Soviet medical administration, these organizations would be designed to provide free medical care to all, and they would do so under the direct control of local soviets without benefit of central administrative direction or of direct administrative control by medical professionals. The articles stressed that these divisions would operate entirely under the authority of the relevant local soviet—that is, under the control of local politicians rather than of medical professionals. As an example the journal cited the Petrograd Medico-Sanitary Division of the Petrograd Soviet, perhaps because of its exemplary structure or, perhaps, because in the chaos of 1918 Petrograd was one of the few soviets with which the editors had regular contact.[48]

The divisions would deal with clinical care, statistics, epidemiology and prophylaxis, forensics, veterinary medicine, health inspection, professional accreditation for physicians and pharmacists, and general administration and finance. They would gather into their framework most of the medical–public health functions—including physician care—in the region under the jurisdiction of a given soviet. Control over medical functions extended to pharmacies and hospitals as well as to pharmacists, physicians, and dentists who were not already employed by a state agency. All health institutions were to be nationalized and individual health practitioners obliged to register, as though for a military draft. Registration forms and the names of registrants were published in the medical journal.[49]

According to G. Karanovich, who should be regarded as a kind of court historian to the Commissariat of Health, a principal difficulty with the framework of local-regional control of medical resources through the soviets in 1917–18 was the lack of local funds to support medical-sanitary activities.[50] Given the economic chaos of the time, which Keep characterizes as a return to primitive barter, financial de-

47. Potulov, *Lenin*, 81–82.
48. *Izvestiia sovetskoi meditsiny*, 25 August 1918.
49. Ibid., 10 October 1918, pp. 2, 6.
50. G. Karanovich, "Etapy razvitiia mestnykh organov zdravookhraneniia" *Biulleten' NARKOMZDRAVa RSFSR* 20 (October 1927), 16–17.

The First Structural Transformation

ficiency seems likely to have combined with problems of control by nonprofessionals to help account for the worsening public health picture.[51] Events, that is, seem to have led to the next phase in the administrative development of Soviet health administration. On 11 July 1918 the new People's Commissariat for the Preservation of Public Health was formally established, and one can say that at the same time Soviet society reached the beginning of the end both of prerevolutionary administrative patterns and of the early postrevolutionary style of administration under soviet executives, the *ispolkoms*.

What "events" appear to have demanded more aggressive administrative measures? The Russian Revolution of 1917 is associated with one of the great demographic disasters of the twentieth century. In a 1922 report to the Health Committee of the League of Nations, L. A. Tarassevitch described a disaster of vast proportions: a legacy in Russia of the worst health of any major European country before 1914 had laid the foundations for the public health debacles of World War I and the revolutionary era.[52] Because of the social and economic upheaval occasioned by the war itself, the Revolution of 1917, and the ensuing civil war, the morbidity and death caused by communicable diseases increased relentlessly throughout Russia. Diphtheria, influenza, typhoid, and typhus fever took a ghastly toll. Typhus, a disease common among large, uprooted populations, spread as a direct consequence of inadequate facilities for simple washing. Beginning in 1918, Russia experienced what was probably the greatest typhus fever epidemic in world history. It was this epidemic that occasioned Lenin's aphorism about the battle between the louse and socialism. And it was these conditions that galvanized the Council of People's Commissars to create in the summer of 1918 a full-blown health commissariat.

The first Commissar of People's Health was the physician N. A. Semashko, a long-time political associate of Lenin. The politics of Semashko's appointment are described by Rigby.[53] Semashko emerges as someone outside the administrative elite or sub-elite that participated in the bureaucracy before 1917. Perhaps he was a "Red Specialist," someone with special technical qualifications who had no commitments to the Old Regime or anyone associated with it and whose future was dependent on the success of the revolution.

51. Keep, introduction to *Debate on Soviet Power*, 16.
52. League of Nations, Health Section, Epidemiological Intelligence, *Epidemics in Russia Since 1914: Report to the Health Committee of the League of Nations by Professor L. Tarassevitch (Moscow)*, pt. 1, *Typhus—Relapsing Fever—Smallpox*, no. 2 (Geneva, 1922), 3–41. For the pre-1914 data, see Ministry of Internal Affairs, Medical Department, *Otchet o sostoianii narodnogo zdraviia i organizatsii vrachebnoi pomoshchi v Rossii za 1907 god* (St. Petersburg, 1909).
53. Rigby, *Lenin's Government*, 131–33.

Z. P. Soloviev, the commissar's deputy, was also a physician but one with impeccable ties to the "public physicians" of the prerevolutionary Pirogov Society. Soloviev was not merely a sympathetic member of the society but he had been, until not long before, an editor of its official journal, *Obshchestvennyi vrach* (The social physician). One would not call Soloviev a "physician-bureaucrat," perhaps, because his professional background was not associated with the major government administrations; undoubtedly, however, he was a "bourgeois specialist" with the kind of professional credentials that marked him as a member of the professional specialist sub-elites that were beginning to emerge in postrevolutionary Russia.

By August 1918 the Soviet medical administration consisted of the central Commissariat of Health and a growing regional-local network made up of the medical-sanitary divisions of local soviets. What had emerged was a central administration dominated by physicians and a regional-local apparatus composed of medical and public health professionals and semiprofessionals. Between these two embryonic structures of a national health administration resided the regional and local soviets and their executive committees, which were beyond the control of the professionals. In order to realize the objective of centralization and its accompaniment, control of professionals, it was necessary for the Commissariat of Health both to gather under its supervision the medical administrations of other commissariats (Education, Labor, Welfare) and to supplant to some degree the authority of the provincial and local soviets.

In late summer of 1918 the official press still wrote of the responsibility of medical-sanitary divisions to local soviets.[54] But by early 1919 numerous communications spoke directly to the local divisions. The registration of physicians in 1919, for example, was required of the local medico-sanitary divisions by the commissariat without reference to an intermediate role by the soviets.[55]

Similarly, as the commissariat gained increasingly large credits for its campaigns against cholera and typhus fever, it became more exacting about how the money was to be used by the local divisions. In the decree "On the Establishment of Funds for Anti-Epidemic Measures," rules were laid down for the acquisition, expenditure, and auditing of funds granted to the divisions for epidemic control. Although an independent Control Commission retained the right to audit expenditures and the local *ispolkoms* retained oversight, the money was explicitly

---

54. *Izvetsiia sovetskoi meditsiny*, 25 August 1918, p. 2.
55. Ibid., 1 February 1919, pp. 3–5.

managed by the commissariat and allocated directly to the local divisions.[56]

In 1919 the commissariat introduced uniformity into the structure of local and regional medical-sanitary organizations. G. Karanovich, writing in the commissariat's official bulletin, identified with remarkable candor the resources with which the it achieved control: "In 1919 with the full concentration of administrative authority, of medical, general, and financial supply, in the hands of the commissariat, the forms of provincial and city divisions of health preservation began to take on a more concrete content. Administration was sharply differentiated from the specialized features of health maintenance."[57]

In other words, the commissariat tried to use both its monopoly over health delivery and its control over financial resources to integrate regional and local administrations into its national apparatus. The success of this effort in terms of public health was no doubt marginal at best. Russia was in the grip of not one but several powerful public health crises, and mere bureaucratic reorganization would not soon change that. Moreover, throughout the 1920s and even beyond, the question of how much centralization was desirable would come up again and again for evaluation. As Christopher Davis shows, for example, during the era of the New Economic Policy (NEP) the central government relinquished the financial club as an instrument of central control, perhaps to the further detriment of public health. Localities were given the responsibility for funding certain kinds of health care, with the result that the funding per capita declined.[58]

By the early 1920s, as Figure 3 illustrates, the state had extensively reorganized medical administration along lines that could only have been determined by medical professional needs. The standardized personnel allocations (shtaty) published in the early 1920s show organizations which, in stark contrast to the prerevolutionary structures, appear to have sharply separated administrative and housekeeping functions from medical-sanitary functions.[59] In addition, these new

56. Ibid., p. 7.
57. Karanovich, "Etapy razvitiia," 15–16. Also see Rigby, *Lenin's Government*, 170ff.
58. Christopher Davis, "Economic Problems of the Soviet Health Service, 1917–1930," *Soviet Studies* 35, no. 3 (1983), 343–61. The Commissariat of Education (NARKOMPROS) engineered a parallel "ingathering" of educational agencies from former tsarist ministries and new commissariats until all but military education institutions were grouped under its authority, according to Sheila Fitzpatrick, *Education and Social Mobility in the Soviet Union, 1921–1934* (Cambridge, 1979), 44, 48–54. The process was not accomplished without difficulty, and under NEP a financial disaggregation occurred in the form of locally gathered fees for education.
59. *Biulleten' NARKOMZDRAVa*, 1 February 1922, pp. 8–9.

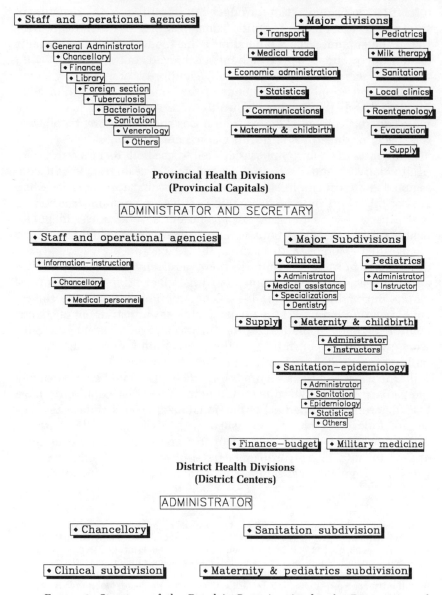

FIGURE 3. Structure of the People's Commissariat for the Preservation of Health, 1922 (RFSFR offices, Moscow). Sources: Central offices, *Vsia Moskva* (Moscow, 1923), cols. 57–62; provincial and district divisions, *Biulleten' NARKOMZDRAVa* 20 (October 1927).

shtaty divided the operational sections of the medical-sanitary divisions along lines specific to medical specializations. This logic of structural development was followed in both central and local administrations. Figure 3 shows an organization whose subdivisions were apparently determined both by the nature of medical practice and the specific perceived health needs of the society.

Determinants of Central Structure

The extraordinary health needs of the society—constituting a situation extrinsic to the political decision-making apparatus—continued to make their demands and to impress political leaders as rationales for administrative development. For example, the commissariat's demands for the registration of physicians and for a redoubled effort to check the emigration of physicians were typically accompanied by references to needs imposed by the cholera epidemic.[60] By the same token, the Council of People's Commissars granted substantial budget increments to the new commissariat for combatting specific epidemics. Special allocations were also made for cleanup campaigns designed to sanitize given areas.[61]

By reflecting briefly on the organizational and political environment out of which the People's Commissariat for the Preservation of Public Health emerged, one can grasp better the degree to which the new administration constituted a departure from the old. Neither during the tsarist era nor under the Provisional Government was *medical* administration unified or centralized. It was simply not conceived of in that way. As we noted earlier, the academician G. E. Rein attempted such a unification almost at the very moment of the demise of the Old Regime, but it never became operational; health administration was never organized along functional lines specified by professional medical practice or demanded by society's medical needs.

On the contrary, other administrative entities consistently subordinated health administration to their nonmedical needs and administrative structures. From a social or political point of view, such structures are not necessarily good or bad. From the viewpoint of the medical profession, however, they carried with them certain distinct disadvantages, the most important of which was that medical practice was controlled both locally and centrally by non-professionals in a way that limited the influence and the career prospects of the professionals.

The Commissariat of Health of 1918 did not achieve a dramatic turn-

60. *Izvestiia sovetskoi meditsiny*, 10 October 1918, p. 56.
61. Letter to N. A. Semashko, 24 October 1921, in Lenin, *PSS*.

around of health care in Soviet Russia, but initially it did try to move in the direction of the Leninist ideal: competent medical care free to all. Even that change, however, was not permanent, as the introduction of health insurance schemes and special health facilities for specific segments of the population would later show.[62]

For my purposes, the most important facts about the emergence of the Commissariat of Health do not bear on whether it was operationally successful. It is far more important that the new administration did not assume the organizational form one would have predicted either from prerevolutionary Party literature or from the diffuse structure of the prerevolutionary health administration. The new commissariat's organizational structure seemed designed both to answer specific severe demands of the moment and to control the application of medical technology throughout the society in the longer term. These functions would be discharged by experts through the application of a specific body of expertise. Even though during the 1920s the relation between the center and localities would be substantially modified, NARKOMZDRAV managed to retain most of the agencies specializing in health care which it had acquired from other commissariats in the early days of its formation, thereby keeping its technical monopoly.

Although pre-Soviet administration relied heavily on professional and technically specialized roles in government, these functions were divided among many governmental authorities. Resources, especially in the medical area, were inadequate; and the control of resources, together with administrative functions, were in the hands of nonprofessionals, often lawyers or bureaucrats without specialization. Each of the principal characteristics of prerevolutionary administration was related to the overall structure of the government—highly centralist, highly authoritarian, but not specialist. Within this framework technically specialized functions were subordinate to the concerns of general, nonspecialist administrations. The controlling role of globally defined governmental administration was of fundamental importance. Thus, while government played a major role in so technical an area as medicine, the functions were subdivided among many agencies, because the generalist administrator considers the specialist simply someone to be used as necessity arises, without regard for the unique characteristics of the specialization.

As the work of John Armstrong and the preceding chapter of this book show, specialists or experts typically hold humble positions in generalist bureaucracies.[63] The tendency in the prerevolutionary Rus-

---

62. Davis, "Economic Problems."
63. John A. Armstrong, *The European Administrative Elite* (Princeton, N.J., 1973),

sian civil administration was for socially elite generalists to occupy powerful and responsible positions in the bureaucracy, while experts who were necessary to organizational operations were inferior both organizationally and socially. Control over policy and administrative structure by nonspecialists naturally may be presumed to have an effect on the specialists' day-to-day activities, which the generalist manipulates in the same way as he handles tenure, promotion, and budgetary resources.

Of course, the revolutionary transformation of administration consisted in much more than a realignment of organizational structure. In the postrevolutionary period immediate action was essential in many areas of government, and this action had to be taken by real people with preexisting class, educational, and career experiences and preestablished values and prejudices. Because of class bias, one may hypothesize, the Bolshevik government found itself unable to use the old elite to discharge administrative responsibilities—but for practical purposes it was unable to dispense with their functions. As a consequence, government seems likely to have turned to other, more palatable reservoirs of the needed skills, creating in effect a transitional administrative corps, which—because of its background and developing organizational status—I call a "sub-elite." As we shall see in the next chapter, not only did this emergent sub-elite find itself in an organizational environment more friendly to specialists than the prerevolutionary bureaucracy had been, but because of the changes in administrative structure just examined in this chapter, the willingness of these specialists to serve the new regime remarkably enhanced their own opportunities for professional advancement.

---

chap. 9; Kendall E. Bailes, *Technology and Society under Lenin and Stalin: Origins of the Soviet Technical Intelligentisia, 1917–1941* (Princeton, N.J., 1978), 31–42.

CHAPTER 4

The Transformation of Personnel in Central Government, 1917–1923

Who were the senior central administrators of the immediate postrevolutionary era? Were they different from their prerevolutionary counterparts? Soviet scholars point out that turnover was very high between 1917 and 1923.[1] In a direct comparison between the incumbents in *high* office of, say, the Ministry of Finance of 1917 and the Commissariat of Finance in 1921, it is not surprising to find continuity among personnel somewhat limited. Accordingly, at first glance N. G. Freiberg, a consultant in medico-sanitary affairs, seems to have been unique in the Commissariat of Health as an individual who held high offices of a similar type both before and after the revolution. But first glances suggesting continuity can be deceiving. Tracing the origins of the officials who were employed in managerial positions in the early 1920s clearly requires doing something more energetic and exacting than simply consulting lists of employees.

This chapter looks at the trans-revolutionary change in the administrative staffs of several ministries commissariats—just as the preceding chapter surveyed the structural changes in several administrative areas of the central government—and shows that the structural changes at the center which broke administration down by functional special-

---

1. V. I. Vasiaev, V. Z. Drobizhev, L. V. Zaks, E. I. Pivovar, V. A. Ustinov, and T. A. Ushakova, *Dannye perepisi sluzhashchikh 1922 g.: O Sostave Kadrov NARKOMATov RSFSR* (Moscow, 1972), chap. 2; E. G. Gimpel'son, *Rabochii klass v upravlenii Sovetskim gosudarstvom* (Moscow, 1982), 170ff.

izations are mirrored in the assignment of appropriately specialized staff. In effect, one witnesses the emergence of an early version of Soviet technocracy, one that continued to evolve. Ultimately, politics came to play a systematic and dominant role even as technocratic functions were extended and intensified.

This chapter picks up the channel of concentration from Chapter 3, beginning with an analysis of the administrative staff of the institutions that became the Commissariat of Health after 1918. Some comparisons between a group of early postrevolutionary incumbents and their tsarist predecessors set the stage for broader comparisons, first with the staffs of other commissariats and then with the central administrative staffs of the Russian Republic in the early 1920s.

The data do not make this task especially easy, but my objective here is to be as precise and comprehensive as possible. After all, we are confronting one of the most lively and pressing problems of the postrevolutionary era in Russia: who, in fact, was running things? The workers? The rich peasants (*kulaks*)? The vanguard of the proletariat— that is, the All-Russian Communist Party (Bolsheviks)? Or as Nicolai Bukharin and, in fits of frustration and paranoia, even Lenin darkly suggested, tsarist bureaucrats—newly converted democrats, born-again socialists who, having turned their coats once, could easily do so again?

If, as I have asserted previously, the organization is mainly its people, and if continuity and operational substance depend on the continuity of personnel, then the revolution could hardly have confronted a more vexing problem: get rid of all the holdovers and you've destroyed the organization; keep them and you've destroyed the revolution. Why? Because, while *we* may have had to prove by tortured examination of statistics that the social forces dominating the tsarist administrative elites were among the most conservative and elitist in the country, that fact must surely have been self-evident to the revolutionaries who lived under their executive authority. And even if you retain only those of the old employees (M. P. Iroshnikov called them *staro-sluzhashchie* [old functionaries]; a 1918 issue of *Kommunist* styled them "public without shame, prepared to serve any master")[2] regarded as essential to operational integrity, you may still risk being charged with subversion in the conditions of civil war and economic crisis that were endemic between 1917 and the time of Lenin's death in 1924. The problem was not only critical but terribly complex; moreover, solutions were comparably complex, pragmatic, and not altogether predictable either from prerevolutionary custom or Bolshevik revolutionary principles.

---

2. *Kommunist*, no. 4 (June 1918), 6–8, quoted in Marc Ferro, *Des Soviets au communisme bureaucratique* (Paris, 1980), 132.

There was no single solution to the problem of staffing the executive arms of the revolution either in 1917, in 1923, or later in the 1920s. Several solutions evolved, matured, and dissipated in a remarkably short time. Moreover, the transformation of staff in the capital masked the truly dramatic transformation in the provinces, where Russia's vast agricultural hinterlands once again led the way to a new era. But that is a matter for the following chapter. Let us now examine the change that revolution wrought in one body of officials, the administrators of government health policies. I begin with a brief characterization of health administration before 1917, after which some direct comparisons with the early 1920s will be possible.

Physician-Bureaucrats before 1917

The key operative or "agent" of the old Medical Department of the prerevolutionary Ministry of Internal Affairs was someone who could be called a "physician-bureaucrat": both a purveyor of medical care to the needy public and a participant in at least some of the administration of health. The status of physician-bureaucrats in the prerevolutionary era was subject to ambiguities, described elsewhere, that were similar to the status ambiguities of other professional-technical groups in tsarist administration.[3] In an era when their numbers and importance to an urbanizing and industrializing society were growing, these professionals, trained as well as or better than any of their bureaucratic colleagues in the central or provincial administrations, were faced with dismal prospects within the organizations that employed them.

Examination of the lists of ranking officials in the Ministry of Internal Affairs and other agencies reveals that there were few prospects for advancement. Rarely does one find the name of one of the thousands of physicians employed by the Imperial government among the top officials of the country. This observation corresponds to the findings discussed in Chapter 2, that individuals trained as nonlegal professionals (engineers or physicians, for example) would only rarely attain elite positions in domestic administration.

With the partial exceptions of the Ministry of Ways of Communication and the relatively subordinate medical administrations of the Ministry of Internal Affairs, there were virtually no cases of accession by specialists even to less elite positions, such as department head, in domestic administration. This meant that professionally trained specialists not only faced poor prospects within the organizations that

---

3. Peter Francis Krug, "Russian Public Physicians and Revolution: The Pirogov Society, 1917–1920" (Ph.D. diss., University of Wisconsin, 1979), chaps. 1–2.

employed them but also exercised little control over the substance of their organizational life—budgets and staffing—and had little opportunity to influence policy decisions. As Peter Krug has shown, they were the object of scorn to their colleagues outside of government.[4] In short, because of their organizational circumstances, physicians played what Rigby calls the game of bureaucratic politics at a considerable disadvantage.[5]

And yet the circumstances under which bureaucratic roles were played out had been changing rapidly. For example, the Medical Department's reports indicate a 223 percent increase in overall expenditures for medical care, with expenditure for hospitals increasing by 150 percent and those for physicians also by 239 percent between 1891 and 1907.[6] The actual number of physicians in all kinds of civil employment was also increasing, according to the 1907 report, in spite of—or perhaps as a perverse consequence of—the fiscal and physical drain caused by the Russo-Japanese War.[7] In short, insofar as the actual functions of health care were concerned, the roles, the importance, and the numbers of individuals specially trained to deal with problems of health in the Russian Empire increased during the last generation of tsarist power, as did the roles, the importance, and the numbers of many other professional-specialist groups.

Nevertheless, if we consider the example of the central Ministry of Internal Affairs, in which one-third of the higher civil servants were statisticians, engineers, and physicians, it is striking that only a tiny fraction of these individuals rose from the sub-elite of section or department head into genuinely elite positions. This fact is accounted for by the characteristics of tsarist administration that we have already seen. Although the ministry was outwardly committed to economic and social development, the personal fabric of its administration was shot through with remnants of an era when land and the patent of nobility counted for everything. Successful careers, it was true, were the product of a good education and long-term commitment to the service. It was equally true, however, that the most successful careers obviously combined these characteristics with good social background and the connections that went with it. Following the convoluted and highly nuanced system of advancement that H. A. Bennett describes, the career of a man with the right background and connections could

---

4. Ibid., 54.
5. T. H. Rigby, *Lenin's Government* (Cambridge, 1979), 116–19.
6. Ministry of Internal Affairs, Medical Department, *Otchet Meditsinskogo Departamenta Ministerstva Vnutrennikh Del za 1891 god.* (St. Petersburg, 1891), and *Otchet ... za 1907 god.*
7. Ministry of Internal Affairs, *Otchet ... za 1907 god*, chap. 6.

advance rapidly in a way that could not have been predicted on the basis of the rules of advancement as outlined in the Imperial code of laws.[8]

The distinction between careers that were simply clean, more or less respectable indoor employment and careers that were exceptionally successful—"brilliant" in the phrase so often applied to them by the novelist Leo Tolstoy—is important to our understanding of the broad social and administrative changes that occurred in the wake of the revolution. In order to develop a better understanding of the difference between a highly successful (and highly unusual, since the man was a medical professional) career and more ordinary ones, let us look briefly at the biography of a man who did achieve to elite status, Dr. G. E. Rein, chairman of the Medical Council of the Ministry of Internal Affairs. Rein was the founder and first head of the tsarist attempt to form an administratively unified health administration. As such, he was obviously a man with political credentials that were as good as his credentials as a bureaucrat. His career was that of an unusually successful physician-bureaucrat.

According to Rein's service record, he was the son of a petty officer.[9] This social background was not, of course, nearly as promising as that of hereditary noble. Unfortunately, although Rein wrote a two-volume autobiography, *Iz perezhitogo, 1907–1918*, the weighty and somewhat self-appreciative work offers little hint as to what, in his early career as a student and young physician, made up for his apparent social disabilities.

Dr. Rein entered active state service in 1875. A graduate of the prestigious Medical-Surgical Academy, he was one of a minority of Russian medical professionals who held the degree of doctor of medicine. His earliest appointments in state service were with the military, but by 1883 he had been named to a university teaching position, and by 1900 he was a professor at the Imperial Military Medicine Academy. It was in 1908 that he was appointed president of the Medical Council, a body responsible for policy formation and administrative oversight. Given the disabilities under which professionally and technically trained civil servants operated, this presidency was perhaps the most prestigious position to which he could aspire in the civil service.

Rein seems to have sought the traditional marks of elite social status in Russian society just as eagerly and successfully as he pursued high administrative office. According to his service record, he acquired his

8. Helju A. Bennett, "Chiny, Ordena, and Officialdom," in W. M. Pintner and D. K. Rowney, eds., *Russian Officialdom* (Chapel Hill, N.C., 1980), 165–68.
9. Ministry of Internal Affairs, *Spisok vysshikh chinov Ministerstva Vnutrennikh Del na 1914 g.* pt. 1 (St. Petersburg, 1914), 84.

own house in Petersburg as well as more than 4,000 *desiatins* of land in Podoliia and Volyniia provinces. He held the title of Honorary Justice of the Peace in the latter (an honorific common to the landed nobility), and in 1910 he was elected to represent the province in the national legislature, the Duma.

In 1912 Rein was surrounded on the Medical Council by individuals not unlike himself—that is, mainly physicians, but also veterinarians and chemists who had seen long, active service in state institutions and who had reached what they must have regarded as the zeniths of their careers, a seat on the council. As a group, they had held their positions on the council longer than Rein himself; some, appointed in the 1890s or even the 1880s, had been there much longer.

In addition to L. N. Malinovskii, the chief medical inspector, the council included Malinovskii's two assistants, the chief health inspectors of the army and the navy, plus the president of the Institute for Experimental Medicine and others of similar background. In addition, there were non-medical representatives from other ministries, such as Justice and Foreign Affairs, where they held active positions. Still, most members were trained medical professionals, although they probably saw few patients.

Salary figures are indicative of the kind of career involved here. Although the council members often held joint appointments in universities or other governmental agencies and although they all held high civil service rank (half had attained rank 3, privy counselor), their mean annual salary was less than the average salary of higher civil servants in the Ministry of Internal Affairs: Rs 5,762 compared to about Rs 6,500 (as Table 12 shows, however, the mean compensation for *council administrators only* was higher—Rs 8,336).

In 1912 the council consisted of about forty members. It was divided into four sections, defined by the nature of appointments. Ten civil servants served in category I as "ex officio–permanent" members; seven served in category II as "representatives of other ministries—permanent"; nineteen held "consultative" appointments in category III; and the remainder served in category IV, the council's own chancellery.[10] Given the distribution of consultative, policy-making, and administrative roles within the council, the members who most clearly fit the classification of "administrator" were those found in categories I, II, and IV.

In addition to those assigned to officials on the council, important medical administrative activities were concentrated in three other agen-

---

10. Ministry of Internal Affairs, *Spisok lits sluzhashchikh po vedomstvu Ministerstva Vnutrennikh Del na 1912 g.* (St. Petersburg, 1912).

TABLE 12.
Status of health/medical administrators in central agencies, 1912

	Medical Council (excluding consultants)	Main Administration for Affairs of Local Economy	Main Physicians' Inspectorate	Institute of Experimental Medicine
Median rank (2 = highest)	3 (privy counselor)	7 (court counselor)	5 (state counselor)	6 (collegial counselor)
Annual compensation (rubles)	8,336	2,769	3,213	2,797
Entry into service (mean yr)	1879	1899	1890	1893
Doctors of medicine (%)	72	0	42	61
N	18	14	12	13
Unknown	3	0	1	4

Source: *Spisok lits sluzhashchikh po vedomstvu Ministerstva Vnutrennikh Del na 1912 g* (St. Petersburg, 1912).
Note: All figures not otherwise indicated are means.

cies in 1912: the Division of People's Health and Welfare of the Main Administration for Affairs of Local Economy, the Main Physicians' Inspectorate, and the Institute of Experimental Medicine. The first two, created in 1904, had direct administrative responsibilities: overseeing hospitals and pharmacies, monitoring medical practice and the operation of certain medical organizations.[11] The third, a research organization, possessed a small administrative staff that was mainly but not exclusively responsible for the internal workings of the institute.

These administrators, who could be thought of as senior and middle-level bureaucrats, totaled some sixty-five individuals in 1912. Given the way status was established in the Russian bureaucracy, it is clear that the Medical Council was by far the most prestigious of the four organizations. That is, in terms of seniority, rank, compensation, and the proportion of officials who held the highest relevant academic degree (Doctor of Medicine), the membership of the council was clearly superior to all the other organizations. Unquestionably, it was the intention of the Minister of Internal Affairs, who was responsible for these appointments, to make the attainment of a seat on the council a high point in the career of a medical specialist.

11. PSZ, 1904, no. 24253.

It is also important to recall that appointees to the council were individuals with long careers in government service: that is, in an institution or agency directly controlled by a central state ministry. Rarely were they physicians with experience in the zemstvo organizations or in city medical institutions; never had their careers been devoted to private practice or to work in industry. In sum, these were doctors who had had extensive administrative, teaching, and research experience but who shared a body of experience that would easily qualify them as both highly successful bureaucrats and physicians.

Not surprisingly, when in 1916 Rein began to staff the short-lived chief administration that he hoped would unify the public health functions of several ministries, he selected individuals with career characteristics similar to those of the members of the Medical Council. Although Rein noted that in addition to medical professionals there were "jurists, persons active in rural and urban affairs, and engineer-specialists in sanitary technical matters," the principal officials were persons of medical-administrative background: administrators from medical schools, members from the Interior Ministry, and the like.[12]

Comparisons between members of the Medical Council and officials in the other three organizations listed in Table 12 are of considerable interest. In particular, they suggest much about the aggregate status of physician-bureaucrats in the Ministry of Internal Affairs, an organization run by lawyers or officials without professional specialization. The differences in salary and other marks of status have already been noted. But it must be kept in mind that in the Russian bureaucracy before 1917 the most consistent determinant of overall status was seniority—that is, time in office, time in rank. As both Bennett and Orlovsky have pointed out, the relationship between status and seniority was complex in terms of both its legal definition and the way it actually took effect within the higher civil service.[13] Nevertheless, once a civil servant had attained rank 7 or 6—especially if he had done so before middle age, thereby demonstrating that he had the right connections—the relationship between seniority and rank became very close. For the Main Administration for Affairs of Local Economy, the Physicians' Inspectorate, and the Experimental Medicine Institute, the zero-order correlations between rank and date of entry were, respectively, .93, .91, and .78. In a sense, because of the uniformity with which seniority cut across the services, other (more subtle) distinctions such as landholding

---

12. G. E. Rein, *Iz perezhitogo, 1907–1918* (Berlin, 1937), 1:97–98.
13. Bennet, "Chiny, Ordena, and Officialdom," and Daniel T. Orlovsky, "High Officials in the Ministry of Internal Affairs, 1855–1881," in Pinter and Rowney, *Russian Officialdom*, chaps. 7, 10.

and class were important, even in the early twentieth century, in marking off differences among officials.

The contrasts in relative status that emerge from Table 12 are subtle but interesting. The health administrators of the Main Administration for Affairs of Local Economy appear, at first glance, to be the most poorly compensated among the officials of four agencies. In fact, considering their junior personal and organizational status, the compensation was not meager but relatively generous. This accords with previous observations, since the Main Administration was the only agency of the four in which there were no physicians—even among those individuals responsible for "hospital administration" and "physician-sanitary administration."[14]

In sum, medical-health administrators in the tsarist bureaucracy, like other specialists, were organizationally diffused and generally inferior in status to their nonspecialist colleagues. By the same token, the personal and organizational status they were obliged to accept assured at best their membership in a sub-elite, rarely if ever in the highest elite.

Health Personnel after 1917

The 1912 central health administration, as we have seen, was divided among four large central agencies that were ultimately controlled by lawyers and nonspecialist bureaucrats. As noted in Chapter 3, the public health and medical administrators from the early 1920s on were concentrated, by contrast, into a single commissariat under the control of physicians. Although officials with careers as physician-bureaucrats were important in both cohorts, their dominance was more comprehensive, and therefore more likely to be effective, among the postrevolutionary group.

Data from 1923, described below, suggest that the early administrators of the Commissariat of Health were about the same average age as the 1912 cohort, but the dispersion of the prerevolutionary group's age and organizational seniority was somewhat greater. As precursor to future developments in an increasingly feminized Soviet medical profession, it is of interest that there were four women among the 1923 bureaucrat-physicians, occupying positions of some significance. (In fact, there had been two women in the 1912 group, but even though they were both graduates of the University of Bern (Switzerland) medical school, they served as laboratory assistants and were not accorded official rank.) A sign of the direction that administration in general was taking in the

---

14. Ministry of Internal Affairs, *Spisok lits sluzhashchikh MVD.*

1920s was the larger number of the 1923 cohort (116) as compared with the 1912 group (65).

Apart from Dr. N. G. Freiberg, who had been a third-class clerk in the Sanitation Administration under the Chief Physician Inspector, there was no one among the 1923 health administrators who had held a central administrative office in 1912. This does not mean, however, that there were no other holdovers from prerevolutionary government positions. A substantial group in the 1923 Health Commissariat in other agencies had occupied government positions before 1917.

Who were these holdovers? Where did they come from? Many were physicians, rather than the lawyers or bureaucratic generalists who tended to dominate central administration in the prerevolutionary health administration. But were they practitioners who had abandoned their independent positions (and viewpoints)? Were they zemstvo, or public, physicians, members of one of the most vocal reform groups before 1917? I have already cast doubt on the notion that they might have been physician-bureaucrats taken over from the dispersed agencies of the Old Regime. But what about the possibility that many were Party activists simply on the lookout for jobs, or even generalist bureaucrats, nonspecialists in the prerevolutionary tradition?

The data necessary to answer these questions are unfortunately neither easily assembled nor straightforward in import. In fact, until recently it seemed that all but the most general characteristics of the early Soviet administrative cadres were inaccessible. In the early 1970s, however, Soviet scholars unearthed the manuscripts of a 1922 census of officials in Moscow.[15] These data, although useful, are at a level of aggregation that makes discussion of specific groups or individuals impossible. Yet such information *can* be compiled by analyzing and combining data from several different sources.

First, the names and positions of the most important officials in the commissariat may be obtained from the Moscow city directory (*Vsia Moskva*) for 1923, the first postrevolutionary year of publication.[16] Second, the names and positions of these individuals, who were serving during the year 1922, may be matched against the first Soviet edition of the register of medical doctors, which provides biographical data: name, sex, date of birth, date and type of degree, specialization, current (1923) employment.[17] Supplied with these facts, one may identify those individuals who were listed in the last available prerevolutionary

---

15. Vesiaev et al., *Dannye perepisi*.
16. *Vsia Moskva: Adresnaia i spravochnaia kniga na 1923 g.* (Moscow, 1923).
17. People's Commissariat for the Preservation of Health, *Spisok meditsinskikh vrachei SSSR (na 1 ianvaria 1924 g.)* (Moscow, 1925).

professional register, the 1912 list of Russian physicians.[18] The 1912 list contains essentially the same biographical data as the 1923 register but adds the name of the medical school that granted the degree. In addition, of course, the 1912 list makes it possible to determine which individuals among the 1923 cohort were actually employed as physicians in 1912 and what the nature of their employment was. Fourth, the 1923 city directory may also be matched against the 1916 *Adres-Kalendar'*, which lists civil servants holding rank (*chin*) only and is thus a measure of status within job categories.[19]

The characteristics of the new generation of Soviet administrators can be grouped in three categories: (1) the ratio of physicians to nonphysicians and their distribution within the organization as a measure of the "professionalization" of the central commissariat; (2) the 1912 employment patterns of the 1923 cohort (that is, who was working where and in what capacity eleven or twelve years earlier); (3) the proportion of those in the 1923 group who were listed as members of the civil service in any capacity in 1916 (in the *Adres-Kalendar'*) as an indication of status or those employed by the prerevolutionary bureaucracy. Let us consider these categories in order.

First, the question of professionalization. Of the 116 Health Commissariat officials listed in *Vsia Moskva* with their offices, sixty-three (54 percent) were also listed as physicians in the 1923 register of medical doctors. That is, in contrast with the Main Administrations for Affairs of Local Economy in 1912, offices in the Health Commissariat that appear to have required medical expertise (such as those in the Venerological Section), the directorships of all subdivisions of the State Institute for Health Preservation, and so on, were in the hands of physicians. Administrations that included no significant medical or public health component, such as transport and supply, were overwhelmingly likely to be controlled by nonphysicians, with the important exception of the highest administrative levels. The top executives—including commissar, deputy commissar, and the members of the commissariat council—were all physicians.

Second, concentrating now on the physician-bureaucrats, one finds that the vast majority of them were trained before 1917. This is true not so much because the preparation of physicians and other medical personnel declined after the revolution—it didn't, apparently[20]—but

18. Ministry of Internal Affairs, Medical Department, *Rossiiskii meditsinskii spisok: Spisok russkikh vrachei na 1912 g.* (St. Petersburg, 1912).
19. Ruling Senate, *Adres-kalendar': Obshchaia rospis' nachal'stvuiushchikh i prochikh dolzhnostnykh lits po vsem upravleniiam v Rossiiskoi Imperii na 1916 g.* (St. Petersburg, 1916).
20. Commissariat for Health, *Spisok meditsinskikh vrachei*; see also Christopher

TABLE 13.
Professional roles in 1912 of Health Commissariat physicians (Moscow officials) in 1922

	N	%
Civil government	9	23.1
Private practice	6	15.4
Instructors	6	15.4
Zemstvo work	5	12.8
Research	4	10.3
Military	4	10.3
Industry	2	5.1
Other	3	7.7

Source: Ministry of Internal Affairs, Medical Department, *Rossiiskii meditsinskii spisok* (St. Petersburg, 1912).

because the commissariat was staffed with persons who were older than the acquisition of a medical degree in 1918 or later would imply. The median year of birth for these officials was 1878, making the average age of the group as a whole about forty-five. The youngest physicians were about thirty and the oldest about sixty-five. In fact, the median year for achieving the medical degree was 1901. Only eight physicians were awarded their degrees after 1913 and only one after 1917. The four women among the sixty-three physicians were all relatively young (about forty) and accordingly had received their degrees more recently.

Of the total of sixty-three physicians employed by the Health Commissariat in the Moscow headquarters in 1923, thirty-nine (62 percent) can be identified as to place and type of employment before the revolutionary upheaval of 1917. Of the twenty-four for whom no prerevolutionary employment data are available, twelve received their degrees after 1912 (and so would not have been in the 1912 list). The remaining twelve are simply missing from the records. Some of these—like N. A. Semashko, the commmissar—may have been abroad, working for the overthrow of tsarism; others may not have been practicing medicine.

The 1912 employment of the 1923 Soviet health administrators is divided into seven categories in Table 13. Apart from the fact that the largest category is that of individuals employed in civil government agencies (23 percent), one may note that at least 46 percent of the commissariat administrators had been employed by the tsarist state in some capacity or other in 1912. The qualifier "at least" is important

---

Davis, "Economic Problems of the Soviet Health Service," *Soviet Studies* 35, no. 3 (1983), 343–61. Whatever the numbers of new graduates, the quality declined, according to Davis.

because the 15.4 percent listed in the instructors' (university–medical school) category were also state employees, holding positions that were traditionally part of the physician-bureaucrat career; George Rein, the highest health official of the tsarist government just before the revolution, included university lectureships among his career assignments. Including these employees of institutions of higher education brings the proportion of former employees of the tsarist state administration to almost 62 percent. In addition it seems likely that most or all of those in the research category were state employees, since most of the small number of medical research institutes were state operations.

It is difficult to be exact about distinctions between a practicing physician who saw patients regularly and one conforming more closely to what is implied by the term "physician-bureaucrat" and would more likely have been involved in medical administration than in patient care. If there was an inverse relation between seniority and involvement in patient care, then it seems that even some of those who were engaged by civil government, the military, and educational institutions may have been performing clinical work. As a group, those who were institutionally employed were about the same age as or just slightly younger than their ministerial colleagues. In 1912, sixteen of the thirty-seven employed physicians would have been under forty years of age.

Comparatively low status for the group in 1912 is also indicated by a cross-check with the *Adres-Kalendar'* for that year or even for the later year of 1916. These directories listed all those officials—but only those officials—who held rank as a consequence of meeting certain criteria during their careers. At a maximum, the 1916 *Adres-Kalendar'* includes only thirteen of the 119 higher officials—that is, both physicians and nonphysician bureaucrats—listed in the *Moscow* city directory as employed by the commissariat in 1922. This suggests that whatever the details of their prerevolutionary backgrounds, the 1922–23 commissariat staff had not held high-level positions of any kind in the prerevolutionary administration.

Holdovers in the Major Central Commissariats

By turning to the entire list of officials (not just physicians) found in the city directory for 1923, we can compare the pattern of holdovers in the commissariat of Health with the patterns in other commissariats: Ways, Finances, the entirely new Supreme Council of the National Economy (VSNKh), the Commissariat of Education, and finally the People's Commissariat of Posts and Telegraphs. Table 14, for example, shows the proportion of officials in four commissariats who are known to have served in a tsarist administration. Each of these organizations,

TABLE 14.
High commissariat officials (1922–23) employed by the state before 1917

	Commissariat (dates of prerevolutionary data)			
	Health (1912)	Ways (1916)	Finances (1913)	Education (1917)
Employed by the state	10.9% (13)	10.1% (22)	12.1% (33)	9.7% (34)
Zemstvo physician	4.2% (5)	—	—	—
All others	84.9% (101)	89.4% (186)	87.9% (240)	90.3% (317)
N	119	208	273	351

Sources: (Health) Ministry of Internal Affairs, Medical Department, Rossiiskii meditsinskii spisok (St. Petersburg, 1912); (Ways) Ministry of Ways and Communication, Spisok lichnogo sostava Ministerstva Putei Soobshcheniia na 1916 g (St. Petersburg, 1916); (Finances) Ministry of Finances, Spisok lichnogo sostava Ministerstva Finansov na 1913 g (St. Petersburg, 1913); (Education) M. B. Kerim-Markus, "Kadry NARKOMPROSa v pervyi god sovetskogo gosudarstva," Istoricheskie Zapiski Academii Nauk SSSR 101 (1978), 83, 84.

whose operations were not especially novel and whose roles in Soviet society under the New Economic Policy were not especially radical, had direct prerevolutionary institutional forebears. Of the 351 high officials and specialists who held positions in the central education administration, the People's Commissariat of Education (NARKOMPROS), some thirty-four had held government positions before October 1917.

The proportion of holdovers that these data indicate is considerably smaller than other sources have shown. Quoting M. P. Iroshnikov, Gimpel'son writes of "old employees" who accounted for 56.2 percent of the Commissariat of Trade and Industry; 97.5 percent of the Commissariat of Finances; and 48.3 percent of VSNKh (whatever its radical pretensions) within a year or so following the revolution.[21] Moreover, holdovers constituted a high proportion of the various staffs of the Commissariat of Posts and Telegraphs. But the criteria I use to identify holdovers in the following paragraphs are comparatively strict. This analysis checks name by name and office by office throughout the corresponding predecessor agencies—Ministry of Finances for the Commissariat of Finances, for example. The reason for this stricter approach is that I am interested in the personal and career characteristics of holdovers as well as in the plain fact that they were holdovers. Thus, the higher proportions cited by Gimpel'son and Iroshnikov may well be more accurate than the ones shown by my data. In addition, of course,

21. Gimpel'son, Rabochii klass, 174–79.

these figures may conceal other characteristics of importance such as covert hostility to the new regime.[22]

In the case of the Supreme Council for the National Economy (VSNKh), pre- and postrevolutionary comparison is highly problematic because there was no clear prerevolutionary progenitor. The tsarist organization whose functions were closest to the main operational thrust of the new commissariat was the Ministry of Trade and Industry, but its holdovers, if any, would more likely have gone to the Commissariat of Trade than to VSNKh. And in any case, the operations of the old ministry and VSNKh were not very similar. In VSNKh the activities of overseeing the operation of Soviet manufacturing and trade had an explicitly political orientation; moreover, as an organization it aimed to be proletarian from its earliest days and, even under the New Economic Policy, radical.

For example, a study of VSNKh published in 1922 notes that there was a substantial difference in educational background between the administrative specialists in the agency and those whose work focused on the administration of specific industrial operations. The positions of the former tended to be rather traditionally bureaucratic; those of the latter, by design, were positions open to workers in the various manufacturing areas. Among the general administrators, according to a survey in 1921, 77 percent had higher education, whereas among the industrial administrators the figure was put at a more modest 57 percent. Indeed, 23 percent of the industrial administrators had no formal education or only primary education. The explanation for the difference, according to the 1922 study, was to be found "in the absence of workers among the general administrative organs."[23] According to Gimpel'son, the proportion of workers among even the industrial administrations was not high, but it was a matter of continuing concern to the political overlords of VSNKh as well as to a group of commissariats, including Agriculture and even Finances.[24]

Thus, it is no surprise that comparative analysis indicates no holdovers in VSNKh. There are several reasons. Not only was it a new organization and not only do we have difficulty in identifying sources in which to search for holdovers, but among those organizations analyzed here it was certainly the most explicitly Communist, the most antitraditional. As a new and radically Communist organization it car-

22. Pitirim A. Sorokin, *Leaves from a Russian Diary—and Thirty Years After*, rev. ed., (Boston, 1950), 306.
23. Supreme Council of National Economy, *Struktura i sostav organov VSNKh v tsentre i na mestakh v 1921 g.* (Moscow, 1922), 24–25.
24. E. G. Gimpel'son, *Velikii oktiabr' i stanovlenie sovetskoi sistemy upravleniia narodnym khoziaistvom (noiabr' 1917–1920 gg.)* (Moscow, 1977), 281–83.

TABLE 15.
Prerevolutionary records of holdovers, 1922

	Commissariat (date of service record)		
	Health (1912)	Finances (1913)	Ways (1916)
Rank as of record date			
median		8	7
Date of birth			
mean	1878	1873	1876 (est.)
SD		6.9	
Date of service entry			
mean	1901 (est.)	1898	1901
SD		7.2	9.3
Salary (rubles)			
mean			4,883
SD			3,179
N	119	273	208

Sources: See Table 14.

ried relatively little baggage with it when the government moved from Petrograd and the state emerged from the prerevolutionary era.

The data for the commissariats of Finances and of Ways of Communication, however, make it possible to extend the characterization of holdovers. Since the prerevolutionary service lists provide some biographical information about the individuals whose names are included, we can characterize the Finances and Ways holdovers as to education, rank, age, and in some cases salary and class. We can thus offer limited, tentative answers to the intriguing questions: Who were they? How did they, among all the tens of thousands of *chinovniki*, "happen" to survive?

Administrators for a Revolutionary Transition

Table 15 summarizes the basic biographical data available for the holdovers in three commissariats: Health, Ways, and Finances. Although their ages ranged from thirty to sixty-five, members of these groups were on the average in their mid- to late forties. They had received their final education near the turn of the century, entering their first service appointments around 1900. Taken together, the three groups were moderately well educated by prerevolutionary standards. The health officials had been mainly trained in their field without the benefit of medical degrees, while about 55 percent of the holdovers from the Ministry of Finances had completed a higher education, and

68 percent of those from the Ministry of Ways of Communication had done so.

As we have already seen, the physicians who found employment with the Commissariat of Health in the early 1920s tended to come not from the administrative levels of the central health administration before 1917 but from a variety of lower-level and nonadministrative assignments. To some degree they were similar, in this respect, to the holdovers from the Ministry of Finances who found employment in the Commissariat of Finances. The Commissariat of Finances was reputed to be one of the redoubts of the tsarist *chinovniki* in the 1920s, but this may simply have been a fashionable yet safe slander of the day. In fact, the vast majority of those officials for whom there are data came from distinctly humble stations in the pre–1917 administration. Clerks, bookkeepers, and assistant and junior technical managers abound. Department heads, chiefs of main administrations, and members of ministerial councils are distinctly absent.

The difference is owing, in part, to seniority. Members of the 1923 cohort were simply not old enough or organizationally senior enough in, say, 1910 or 1915 to have achieved senior positions. This is confirmed by the median rank of the Ministry of Finances cohort in 1913, which was only collegiate assessor (rank 8), closer to the bottom than the top. Since it appears that rank in the prerevolutionary period was more closely tied to seniority than to any other measure of career progress, the average attainment of rank 8 suggests a junior group.

The average suggests still more. Even the oldest official, a certain Petr Alekseevich Smirnov, who was born in 1860 while Alexander II still reigned, had attained only rank 12, provincial secretary, by 1913. Certainly, Smirnov's slow progress can in part be explained by his education, or lack of it: his published career data list no formal education and indicate that he had been admitted to the distinction of a ranking official only by examination—probably after spending some time as an unranked clerk. Another illustration can be found in the record of Leonard Frantsevich Getlikh, whose church affiliation, in spite of his un-Russian sounding name, was listed as Orthodox. Born in 1867, Getlikh was a graduate of no less an institution than the University of St. Petersburg. Nevertheless, by 1913 he still held only rank 12, provincial secretary. At that rate, he could have looked forward to retirement as a collegiate assessor (rank 8)—with luck.

The picture that emerges, then, is that of a slow-rising, perhaps underprivileged group of bureaucrats whose opportunities seem to have been improved by the revolution. I conclude that many former tsarist officials made the transition to the new finance agency, but they tended to have been young and junior in status before the revolution, at the

The Transformation of Personnel

low end of the social stratum I have called the sub-elite. Quite possibly, such individuals had aspired to advancement in administrative rank, but in prerevolutionary conditions they would have had only minimal chances of promotion. Given prerevolutionary organizational structures and the characteristics of career advancement, it was unlikely that most of these junior technicians would have climbed the ladder very far at all *without the revolution*. In 1913 they were *sub-elite*, not *pre-elite*.

A somewhat different image emerges, however, from close examination of the group of holdovers from the Ministry of Ways. Responsible for maintaining, rebuilding, and developing various communications networks, these officials played a significant role in the survival of Russian railways both before and after 1917; indeed, their functions were essential to the survival of the Russian national economy, and they were no doubt difficult to replace. That, at any rate, is the explanation that presents itself for the extraordinary fact that among the holdovers were a former privy counselor (rank 3), two active state counselors (rank 4), and a state counselor (rank 5), as well as a major general.

Nor can these elevated ranks to be accounted for merely by age. One of the active state counselors had entered state service only in 1894, the state councelor in 1897. These holdovers are examples of the kind of career advancement that was available, under the tsarist system, to individuals who possessed either a truly distinguished educational background or (often the same thing) good social credentials. While class data were not published for the entire holdover cohort, prerevolutionary sources show that three of those who had achieved distinguished rank were either members of the hereditary nobility or children of officials.[25] In 1916, moreover, these individuals boasted both offices and salaries to match their titles. Two of them earned more than 8,000 rubles annually (a lot for the time), while one Petr Aleksandrovich Malevinskii, identified in an 1898 service register as a hereditary nobleman, received the rather princely sum of 15,000 as administrator of the Moscow-Kursk and Nizhnii-Novgorod Railway.[26] Landing solidly on his feet after the revolution, Malevinskii held the position of deputy chief of Ways and Communications in 1923. The other former high officials then in the ministry held equally distinguished position in the early 1920s.

These three men were admittedly not typical of holdovers in general; indeed, they are not even typical of the holdovers from the Ministry of Ways of Communication, as a reexamination of Table 15 will reveal.

---

25. Ministry of Ways of Communication, *Spisok lits sluzhashchikh na 1898 g.* (St. Petersburg, 1898).
26. Ibid.

Still, in their cases, economic and technical necessity seems to have triumphed easily over politics as a determinant of position—at least to a degree. It is worth noting, however, that just as the Moscow city directory lists these high officials in the commissariat, it also lists, as commissar of Ways of Communication, one Felix Edmundovich Dzerzhinskii, who simultaneously held other positions that made him responsible for fighting against counterrevolutionaries, including elite members of the Old Regime. Dzerzhinskii was the leader of the first Soviet political police organization. Maybe he was made commissar to ensure that the blueblood technocrats who managed the commissariat truly served the interests of the new state.

On the basis of these imperfect and admittedly tedious observations we can perhaps conclude that, overall, the revolution did immediately offer opportunities even to former *chinovniki* that they could not reasonably have expected in the absence of such an upheaval. Times were certainly hard; even day-to-day survival was occasionally called into question by war, pestilence, and famine. Nevertheless, these officials, or most of them, had reason to thank some lucky star, perhaps even the ubiquitous red ones that were to be seen everywhere in Moscow.

It is also clear that the revolutionary government was willing to pay, both in terms of administrative prestige and probably in money, for scarce managerial and technical skills. Thus, when it could find bookkeepers or physicians in lowly stations and integrate them into the postrevolutionary commissariats, it did so. But when driven by necessity, it was prepared to offer even distinguished and powerful positions to members of the old elite. Admittedly rare, the instances nevertheless underscore the fact that whatever the preferences of revolutionary politics, society and its economy needed to go on functioning.

A brief examination of the experience of one more commissariat in the middle 1920s, Posts and Telegraphs, requires consideration separate from the one just concluded, because the data are different. First, we have access to an extraordinary resource from 1927, a major survey of some 92,000 employees of the commissariat.[27] Unfortunately, however, no satisfactory record of officeholders in the imperial Main Administration for Posts and Telegraphs from before 1917 has been obtained; therefore direct comparisons of pre- and postrevolutionary officeholders is not possible. Moreover, the 1927 survey of commissariat employees, while comprehensive in many respects, does little to answer *where*, exactly, senior administrators came from.

Table 16 shows the seniority patterns within both the commissariat

27. People's Commissariat of Posts and Telegraphs, *Perepis' rabotnikov sviazi 27 ianvaria 1927 goda* (Moscow, 1929).

The Transformation of Personnel

TABLE 16.
Date of formal appointment for Posts and Telegraphs officials and employees, 1927

	N	To 1889	1890–1913	1914–17	1918–26	Unknown
Commissarial offices	4,260	1.8%	32.3%	15.7%	50.1%	0.1%
Post-telegraph system	70,989	1.0	32.9	26.8	39.1	0.2
Railway agencies	5,854	1.1	36.8	29.6	32.3	0.2
Telephone system	11,185	0.3	19.6	20.5	59.4	0.1
Radio system	688	0.0	7.7	9.2	83.0	0.1

Source: People's Commissariat for Posts and Telegraphs, *Perepis' rabotnikov sviazi, 27 ianvaria 1927 goda* (Moscow, 1929), 41, Table 3.

TABLE 17.
Date of formal appointment for Posts and Telegraphs administrative personnel, 1927

	N	To 1889	1890–1913	1914–17	1918–26	Unknown
Commissarial administrators	7,169	1.8%	60.7%	26.2%	11.3%	0.0%
Agency managers	3,584	1.2	39.8	33.1	25.9	0.3
Technical administrators	621	0.8	53.0	20.0	25.9	0.3

Source: People's Commissariat for Posts and Telegraphs, *Perepis' rabotnikov sviazi, 27 ianvaria 1927 goda* (Moscow, 1929), 42, Table 4.

and its affiliated agencies (post offices and the like) throughout the country. The staff of the commissariat itself was exceptional in that a majority achieved formal appointment after the revolution. A majority of the workers in the telephone and radio systems were also appointed after 1917, but one is tempted to explain this as a consequence of some greater affinity for modern technology among the young. It is important that the numerical majority of the entire apparatus of Posts and Telegraphs, including employees in the postal and railway systems, were prerevolutionary holdovers. It is equally interesting that all groups of administrative personnel—the senior administrators, agency managers, and technical administrators described in Table 17—harbored a majority of holdovers.

In the light of the findings on other commissariats, it is tempting to conclude that these administrators were sub-elite and junior officials from the prerevolutionary era for whom being held over also meant moving up. It is most frustrating, then, that the data do not allow for such a determination. Since we have no information on prerevolutionary status of these officials, we have no grounds for concluding that they were sub-elites who moved into elite position because of the Revolution of 1917. Indeed, the rather high proportions of officials who were appointed during World War I suggests that other factors than those associated with the upheaval of revolution—such as the stresses of the war—accounted for significant turnover. Still, the data from the Posts and Telegraphs survey were not without social and political significance at least to those who gathered them in 1927. V. N. Bobrikova, whose observations on the data appear throughout the published version, frequently comments on the limited opportunities that staff turnover in the system had provided for workers and peasants.[28]

Though the Posts and Telegraphs survey is not as helpful as it might be, the available aggregate data do help to reveal the characteristics of turnover and holdover immediately after 1917. Vasiaev, Drobizhev, and their colleagues remarked on the relatively short tenure of officials appearing in their 1922 Moscow sample: of 3,078 officials in the sample, only 422 (13.7 percent) were employed by their 1922 agencies as early as 1917–18, and some 67.5 percent did not take up their appointments until 1921–22—that is, during the year that preceded the census.[29] One would be justified in thinking that such short tenure simply reflected the upheaval of the times—much movement of population, unexpected deaths, premature resignations—but the data are more than a mere barometer of the tempestuous environment. For one thing, in spite of the short tenure, almost 70 percent of the 1922 cohort had held positions of some kind in the civil service before 1922, and the vast majority of these held high or middle-level positions in the post–1917 administrations. One has the impression of a great deal of movement, perhaps from one government position to another, or from extragovernmental positions to state service.

Table 18 gives some idea of the nature of the movement of 1922 officeholders before they reached their 1922 positions. The "top officials" column (category A in Vasiaev) makes clear that about 30 percent of the 1922 Moscow cohort held official positions of some kind before October 1917. As we have seen, these positions would not likely have been in the higher civil service, because of age or seniority factors and the tendency for such offices to have been dominated by noble and landed social elites. Thus, of the total proportion of 29.7 percent ac-

---

28. Textual commentary by V. N. Bobrikova, ibid., 14–19, 30–36.
29. Vasiaev et al., *Dannye perepisi*, 69.

TABLE 18.
Career activities before and after October 1917 of officials in Moscow commissariats, 1922

Activity	Top officials				Specialists				Junior-lower grades			
	Before Oct. 1917		1917–22		Before Oct. 1917		1917–22		Before Oct. 1917		1917–22	
	N	%	N	%	N	%	N	%	N	%	N	%
Higher-middle state service	757	27.9	1,441	55.9	181	10.9	395	24.7	22	5.1	166	12.7
Lower state service	48	1.8	298	11.6	33	2.0	187	11.7	53	3.5	188	14.4
Students (non-engineers)	437	16.1	36	1.4	419	25.2	90	5.6	58	3.8	48	3.7
Junior technicians	248	9.1	—[a]	—	120	7.2	—[a]	—	170	11.2	—[a]	—
Intelligentsia (including teachers & physicians)	343	12.6	109	4.2	361	21.7	310	19.4	233[b]	15.4	188[b]	14.4
Military	294	10.8	451[c]	17.5	90	5.4	338[c]	21.1	479	31.7	384[c]	29.3
Engineer/tech. work	114	4.2	220	8.5	153	9.2	157	9.8	9	0.6	243	18.6
All others classified	88	3.2	21	0.8	43	2.6	105	6.6	160	10.6	92	7.0
Unclassified or unknown	383	14.1	—	—	214	12.9	18	1.1	273	18.1	—	—
Total	2,712[d]		2,576[d]		1,662		1,600		1,512		1,309[d]	

Source: V. I. Vasiaev et al., *Dannye perepisi sluzhashchikh 1922 g.* (Moscow, 1982), 61, 62, 89, 90, 125.
[a]Included in "Lower state service" category.
[b]*Litsa svobodnykh professii.*
[c]In Red Army after October 1917 (includes prisoners of war).
[d]Total given in source is in error.

counted for by the first two activity categories, it is likely that most would have come from the middle or lower ranks of the prerevolutionary service.

It is noteworthy that the second largest category among top officials giving pre-1917 occupations is that of student. This was presumably an upper-class group (though probably not noble) comprising persons who were quite young at the time of the revolution. One might make the same statement about the military category, not only because there was always a tendency for brief military careers to precede civil careers but because the year 1917 was preceded by two and a half years of war. It is also noteworthy that the other large (12.6 percent) category, the intelligentsia, includes teachers and physicians—upper-class groups, to be sure, but organizationally and socially inferior on the tsarist scale of status.

One should observe further that urban and agricultural workers are not present in significant numbers—a circumstance that was obviously the product of the tsarist education system but may also have been a consequence of workers' attitudes toward upward mobility. If, as Sheila Fitzpatrick suggests, workers were relatively slow to exploit new educational opportunities, one may also surmise that they were relatively slow to exploit opportunities to leave the status of laborer behind for something entirely different.[30]

It appears, then, that the prerevolutionary experience of top officeholders in 1922 was parallel to that which one would associate with 1922 junior (in seniority) sub-elite groups. This said, one must keep in mind that while nobles and landholders were presumably absent by 1922, there is not as yet evidence of any seizure of a significant number of higher offices by the tsarist lower classes.

Nevertheless, the rapidity with which upward mobility occurred for some groups is suggested by the additional data provided for top officials, particularly Vasiaev's data on activity from October 1917 to the time of appointment to the 1922 position. During that period more than two-thirds of the cohort had turned to civil service, with an additional 21 percent occupied in state service of other kinds (military, teaching, or medical care, for example). Moreover, the movement is not simply lateral—that is, from one sector to another of equal status, a great deal of it is upward. The activity category of higher-middle state service is approximately twice the size of the same category before 1917, while the student category has shrunk to less than 10 percent of its earlier

---

30. Sheila Fitzpatrick, *Education and Social Mobility in the Soviet Union* (Cambridge, 1979), 12–17.

The Transformation of Personnel 117

total, and the intelligentsia (teacher/physician) category in 1922 is about one-third its size before 1917.

Additional insight may be derived from comparison of the top officials category with the second and third parts of Table 18: "specialists" (economists, statisticians, engineer-administrators, physician inspectors, and the like) and "junior-lower grades." The proportion of pre-1917 state servitors is much lower in these categories and the proportion of intelligentsia somewhat higher than those found among top officials, suggesting that more moved into the specialist and junior grades from outside the bureaucracy than from within. More important still, movement into the civil service at any level occurred much more slowly for these groups than for 1922 top officials. Part of the difference in mobility was certainly accounted for by the fact that among junior grades the proportion of students was quite low (3.8 percent) before 1917, although perhaps 27 percent of the group would then have been of the correct age to be in school.[31]

In the data on education and age, additional contrasts emerge which seem to confirm the rise of formerly inferior tsarist groups. It should come as no surprise that the specialists were the most highly educated of the three groups. Some 61 percent of this category had some higher education, and an additional 9.5 percent had at least a secondary or "specialist" education.[32] The educational profile of the junior-lower grades is, not unexpectedly, much worse: only 45 percent had some higher education.[33] The top grades were quite close to the specialists in terms of educational experience in 1922, as they had been before the revolution.

The most important implication of the educational data is that the 1922 cohort on the whole was less highly educated and less adequately educated than their tsarist predecessors. This finding, though not unexpected, is significant for several reasons. Given pre-1917 educational patterns, the substantial proportions of the top officials and specialists who had received higher and secondary or specialized education appears to lend credence to the view that these were inferior social classes emerging out of relatively obscure positions into organizationally elite positions.

The age data present few surprises, given the observations already made. The junior-lower category includes a relatively young cohort: approximately 45 percent were under age thirty, as compared to 38 percent of the specialists and 29 percent of the top officials.[34] Again,

31. Pitirim A. Sorokin, *The Sociology of Revolution* (Philadelphia, 1925), 240–43.
32. Vesiaev et al., *Dannye perepisi*, 116–17.
33. Ibid., 85.
34. Ibid., 118.

by comparison with tsarist-era data, these groups seem relatively young. The women who were making their way into senior positions in state service for the first time appear to have been extremely young. The top category includes 270 women, of whom 63 percent were under thirty; the specialist category includes 452 women, of whom 53.5 percent were under thirty; the junior category includes 853 women, of whom 56.7 percent were under thirty.

Impact and Survival of the Holdovers: The Later 1920s

Having identified at least the main characteristics of the central administrative Moscow cohort of 1922, we can inquire whether the generation of civil servants of which it was a part, the first generation to establish itself after the revolution, created a pattern for the future. Did their successors tend to be like them?

In the normal course of events, one would expect to find more similarities than differences. Of all those who are able to influence the formation of a new generation in most organizations, those who control the organization at the moment of entry of a new cohort would appear to be in the strongest position.[35] However, numerous factors were at work nullifying the influence of those who already occupied the high ground of bureaucratic office. The Party, in particular, had every reason to work against the influence of the old bureaucrats.

There were, in effect, halfway houses open to a revolution that wanted to achieve organizational change quickly but also wanted to avoid the catastrophe that out-and-out destruction of a given bureaucracy would produce. Thus, as we saw in Chapter 2, substantial restructuring occurred, a restructuring initiated first by the Provisional Government and then intensified by the Bolsheviks. Old functions were not abandoned, married as they were to a society that demanded they continue, but they were reorganized, regrouped, and reconceptualized.

Nor, as we have seen, did the old tsarist hierarchy manage to hold on in any meaningful sense. By 1922 the central administration was controlled by a generation more or less free of the presence of persons who had held high positions in the tsarist administration. For such a change to have occurred at any time during the nineteenth century would have required a truly revolutionary turnover. Still, the 1922 cohort of top administrators included a majority of persons who had been associated with the prerevolutionary government and its insti-

---

35. It should be noted that given the age profile presented (ibid., 51), the proportion listed as "students" or as "military" seems high.

The Transformation of Personnel

tutions as teachers, physicians, soldiers, students, and of course bureaucrats—sometimes even as high-level bureaucrats if their skills were rare enough. The influence of the last prerevolutionary generation of government servitors was still felt, not as strongly as in the previous generation to be sure, but felt nonetheless. Perhaps Lenin, Trotsky, Bukharin, and all the other Bolsheviks who complained of the presence of too many *chinovniki* were not paranoid but simply trying to face up to a problem that would not go away as fast as they had hoped it would.

How long did the influence of holdovers last? More specifically, was the prerevolutionary generation still present at the end of the 1920s, the first decade of Communist rule and the beginning of the era of intensive industrialization? Although this question anticipates Chapter 5, it is worth considering here.

If one returns, first of all, to examine individuals in specific commissariats, it is evident that the overall turnover noted by Vesiaev and his colleagues was still very much the order of the day. We can evaluate the survival of the cohort of 1922 until the end of the 1920s by checking successive editions of the Moscow city directory. The overall pattern of attrition is substantial by comparison with pre-1917 experience. In the Ministry of Internal Affairs before 1917, turnover occurred at approximately the rate of 10 percent per year among higher civil servants. If one takes this pattern, established between 1905 and 1916, as a standard, then from 1922 to 1929 every one of the commissariats examined exceeded the pre-1917 rate, three of them—Health, Finances, and Agriculture—substantially.

Professional expertise and experience do not seem to have been absolute guarantees against high turnover. In the Health Commissariat, for example, it is evident that physicians too experienced high rates of wastage, though their departure was not as fast as that of their less technically qualified colleagues: by 1927, among the three commissariats, physicians in the Commissariat of Health experienced the least attrition. The importance of professional expertise is further underscored if one compares the experience of the three commissariats with that of the Central Statistical Administration. Here attrition was high by comparison with pre-1917 experience and perhaps very high by comparison with the pre-1917 turnover of professional-technical personnel. Still, compared with data from other commissariats, the turnover data from the Central Statistical Administration clearly suggest a relatively favorable experience for this more highly technical and professional agency.

The pattern of turnover within commissariats is of some interest. It is evident that neither survival nor its opposite, wastage, was a random or even universal phenomenon. Instead, each was associated with specific departments or divisions in a given commissariat. For example, of the entire group of survivors in 1927 in the Health Commissariat, about 30 percent were in one department, General Financial Administration. Two-thirds of the survivors were to be found in only three departments, of which two were medical. In the Central Statistical Administration, holdovers were distributed in a similarly skewed fashion: the three departments with the highest proportion of holdovers (Statistical Exchange, Social Statistics, Administrative-Financial Administration) included about 63 percent of the total of survivors.

Similarly, in the Finances Commissariat survival was associated with departments. For example, the Budget Administration alone accounted for 35 percent of all holdovers, as did the Taxes and State Income Administration, for a combined 70 percent of all holdovers—although these two divisions employed only 47 percent of all officials listed in 1927. The experience of officials in the Agriculture Commissariat is more difficult to characterize. The rate of attrition there—as in VSNKh, a highly politicized body—was much higher than in other commissariats; by 1927 the survivors constituted a very small and dispersed proportion of the total body of officials. Given that the Commissariat of Agriculture was responsible for implementing the politically sensitive policies connected with land reform, there was much to be gained by barring tsarist holdovers from infiltrating its ranks.

Elite Replacement: Pre- and Postrevolutionary

Turnover during the 1920s certainly appears to have been different from turnover before 1917. In addition to such "normal" factors as promotion, transfer, retirement, voluntary resignation, and death, it was clearly influenced by other factors—political considerations and new administrative initiatives—as well. It seems reasonable to suppose that these would account for the differences between the pre- and postrevolutionary experience just as they would explain why the survival of officials in the 1920s was not randomly distributed throughout commissariats but concentrated in specific departments.

Still, it is not necessarily the case that a department head survived together with others in his department; often, only subordinates in a given department were holdovers. If we assume that other factors, apart from characteristics of technical training, were operating to ensure the survival of specific persons, then it may well be that those factors were taken into account when the staff of the department was assembled in the early 1920s.

It should certainly be observed that the period under examination was one when the system of *nomenklatura* was established. Nomenk-

latury consisted of lists of state and party offices over which given party organizations had the power of appointment. As early as 1924, even provincial party organizations exercised this control over extensive ranges of offices at the provincial and district levels of government.[36] The principal objective was to ensure that persons who were appointed were acceptable to the responsible political authorities, whether the appointment was civil, military, or political. This is only one reason why it is obvious that replacement or survival of administrative cadres was affected in the 1920s by factors that are not evident from an examination of biographical and career characteristics alone.

Finally, it should be noted that the entire decade of the 1920s was an era of very broad administrative reorganization. The creation of the *nomenklatura* under the control of the Party secretariat altered radically the relationship between the Party and other administrative agencies, especially when Lenin's departure from the scene is taken into account. At the same time, the Russian Republic was replaced by a three-tier administrative apparatus; shortly, thereafter, provinces were replaced by new administrative entities called *oblasts*. The implications of these changes for the transformation of personnel were very likely considerable.

The problems posed for us by "bureaucratization" and structural change after the Russian Revolution are, of course, anything but new, just as they are anything but simple. Alexander Gershenkron has discussed the complexities of defining continuity and change and especially the difficulty of applying any such concept to a real—that is, historical—social situation.[37] In part, this theoretical or conceptual difficulty is magnified by the way the "historically important" individuals, usually politicians, address themselves to the products of change and especially by their tendency to seize upon one or two singular factors—such as "hangover" of personnel (Lenin), or the restructuring of property ownership—as indicative of broad social continuity or discontinuity.

The "colossal administrative apparatus" that was in place in the early 1920s was a source of concern and virtually constant political debate for leading Soviet politicians of the day who, like Lenin, were embarrassed by its similarity to the prerevolutionary bureaucracy.[38] Bukharin criticized the entire revolution because, "taking too much on itself, it has to create a colossal administrative apparatus. To fulfill the economic functions of the small producers, small peasants, etc., it requires too

---

36. Merle Fainsod, *Smolensk under Soviet Rule* (New York, 1958), 45.
37. Alexander Gershenkron, *Continuity in History and Other Essays* (Cambridge, Mass., 1968), 11–39.
38. Sorokin, *Sociology of Revolution*, 252–53.

many employees and administrators. The attempt to replace all these small figures with state *chinovniki*—call them what you want, in fact they are state *chinovniki*—gives birth to such a colossal apparatus that the expenditure for its maintenance proves to be incomparably more significant than the unproductive costs which derive from the anarchistic condition of small production.[39]

Bukharin's use of the term *chinovniki*, traditionally applied to tsarist bureaucrats, was meant not merely to deride the presence of holdovers but, perhaps more, to assert that all the *new* bureaucrats too were, "call them what you want", *chinovniki*.

In the mid-1920s, according to Isaac Deutscher, Trotsky was also critical of the bureaucracy but more narrowly so, since he had earlier been a strong supporter of administrative development and centralization and since the principal target of his antibureaucratic rhetoric was Stalin's Party secretariat.[40] In fact, it is possible to interpret the struggle for succession to the mantle of Lenin in terms of the attitudes of competing factions toward administration and its role in contemporary and future Soviet society. (I return to the Stalin-Trotsky controversy over this point in Chapter 5's discussion of the emergence of the role of the Party in elite formation in the 1920s.)

In spite of the evident importance of this controversy, it is tempting to assert that the spokesmen for the warring factions might as well have saved their breath and ink. The rhetoric is as deceiving now as it was irrelevant then. The debate implied that by 1923 there was a real choice as to whether the new administration would be sustained. In fact, there was no realistic alternative, and it is interesting that even Lenin, shortly before his death, remarked upon his feeling of powerlessness to steer the vehicle of government in the direction he wanted it to go.[41]

The Revolution of 1917 only made the new administrative system possible. It provided—perhaps required—the formal circumstances necessary for the formal changes of structure that began in 1917 and continued thereafter, "rationalizing" ever wider areas of economic and social life into one administrative niche or another. It also opened the way for the emergence of the pre-1917 technocratic sub-elites into positions of genuine influence. But this is not the same thing as saying that the revolution—or Bolshevism—caused or even created the new system. We must look not merely at the revolutionary year but at the whole revolutionary generation of 1890–1930, the generation that, not

39. Quoted in Stephen Cohen, *Bukharin and the Bolshevik Revolution: A Political Biography, 1888–1938* (New York, 1973), 140.
40. Isaac Deutscher, *The Prophet Unarmed: Trotsky, 1921–1929* (New York, 1965), chap. 2, esp. 119ff.
41. V. I. Lenin, *PSS*, 5th ed. (Moscow, 1958–65), 33:235–76.

coincidentally, lived through both the Bolshevik Revolution and the first great industrial spurt in Russian history. The causes of structural transformation must be sought in the continuity and discontinuity bred by change over the entire course of these modernizing decades.

Most of the thousands of professional-technical officials who staffed the post-1917 administration, who gave government some of its form and most of its technical content, were given their opportunity to advance from humble, subordinate status to the level of elite management by the political-social revolution in 1917. But they were the direct product of the first generation of the Russian industrial revolution. Their technical skills would not have been available to the Revolution of 1917 and the Bolsheviks had it not been for the drive to industrialize and urbanize that occurred in the 1890s and later. Neither would they have possessed the administrative experience that made them useful in 1920 if the industrialization and technical development of the 1890s had not combined with the tsarist compulsion for bureaucratization to force the generalist ministries of the last tsarist era to train these officials and then exploit their skills. The Revolution of 1917 gave them their chance to leave a heavy imprint on the second generation of Russian industrialization through the legacy of a specialist administration.

CHAPTER 5

The Communist Party and the Soviet Administrative Elite, 1922–1930

Among the many Westerners who came to Russia on some mission during World War I and stayed to witness the Revolution of 1917, Pierre Pascal must be counted as one sympathetic to the Bolshevik cause. Partly for this reason he stayed throughout the 1920s. In 1927, despite good political connections, he fell afoul of the Commissariat of Finances and was obliged to endure a kind of tax audit. He recorded in his diary an impression of his visit to the commissariat: "What a vivid picture of bureaucracy, that immense room, thick with white-collar workers seated at rank on rank of tables; at either end, portraits of Marx; on the sides, Stalin, Tomsky, Lenin. Some white-collar workers, young people, had an aura of the worker; others, not at all. Someone was interrogating a photographer about his income and his expenses, his housing, his family, detaining him throughout the entire half-hour I was waiting and even afterward—politely, but in such detail and with such insistence!"[1]

A young man's impatience with the fussy discipline of a bureaucratic environment and bureaucratic behavior comes through clearly. One senses, equally, a revolutionary's disdain for the power of bureaucrats. But I wish to underscore an observation that might otherwise be missed: the "others" had nothing about them that bespoke the working class. It reminds one of the aspersions cast by Bukharin, quoted earlier, against

1. Pierre Pascal, *Russie 1927: Mon journal de Russie* (Paris, 1982).

The Communist Party and the Soviet Elite 125

the new *chinovniki*. Who were these people who earned a living in state service at the end of the 1920s? Had the revolution fallen, after all, into the hands of clerks and lumpen intelligentsia?

This chapter formulates answers to these questions about change in the personnel of the Soviet civil administration during the middle and later years of the 1920s. The changes are described, of course—as are patterns of development undergone by certain segments of the civil administrative apparatus: the nonmilitary domestic commissariats, the Council of People's Commissars, the Central Committee. But they are also treated as the product of interaction among objectives of the political elite on the one hand, and the ambitions of traditionally underprivileged social classes on the other. In short, the main purpose here is to explain change in civil administration as the product of interaction between political elites and upwardly mobile administrative recruits.

Further, although I am very much interested in the problem of political control over bureaucratic development, this chapter does not aim to specify in any detail what Lenin, Stalin, or other political leaders wanted or didn't want out of Soviet state administration during the 1920s; that has already been done by many specialists.[2] Lenin's writings on the subject, moreover—books such as *Will the Bolsheviks Retain State Power?* and *State and Revolution*—are well known.[3] I rely on specialist studies and the broad characterization of Lenin's views as a point of departure in an effort to see whether bureaucracy became, in the 1920s, what the Bolsheviks wanted it to become.

Using the necessary organizational and career data, this chapter outlines specific features of the administrative apparatus and its personnel during the 1920s and discusses the implications of these findings for the problem of upward social mobility during the period. If it is correct to say that one of the objectives of the revolution was to break the traditional shackles that controlled and inhibited the upward movement of inferior classes in tsarist times, then an evaluation of administration as a channel for upward social movement is obviously in order. There was no greater employer in the Russian state, *ancien* or *nouveau*.

---

2. In addition to the major studies of the formation of the early Soviet government cited in Chapters 3 and 4, see Jeremy Azrael, *Managerial Power and Soviet Politics* (Cambridge, Mass., 1966); Richard B. Day, *Leon Trotsky and the Politics of Economic Isolation* (Cambridge, 1973); V. S. Orlov, "V. I. Lenin i nachalo deiatel'nosti Malogo SOVNARKOMa" *Voprosy istorii* 9 (1974), 210–13; D. I. Sol'skii, "NOT i voprosy deloproizvodstva (1918–1924 gg.)," *Sovetskie arkhivy* 6 (1969), 47–52; M. O. Chechik, "Bor'ba kommunisticheskoi partii za ukreplenie i uluchshenie raboty sovetskogo gosudarstvennogo apparata v gody pervoi piatiletki (1928–1932 gg.)," *Vestnik Leningradskogo Universiteta: Seriia istoriia, izayka i literatury* 8 (1958), 49–66; E. G. Gimpel'son, *Velikii oktiabr' i stanovlenie sovetskoi sistemy upravleniia narodnym khoziaistvom* (Moscow, 1977), and *Rabochii klass v upravlenii sovetskim gosudarstvom* (Moscow, 1982).

3. V. I. Lenin, *PSS*, 5th ed. (Moscow, 1962), vols. 33, 34.

Nor was there any more persistent reminder to the ordinary citizen of the power of that state than those inevitable and frequent dealings with the bureaucracy which life in Russia required. If improvement in station and status could not be had by the lower classes through employment in state administration, then it might fairly be argued that the revolution was socially meaningless, whatever its economic or political implications.

The structure of this chapter is straightforward. I first establish what seem to have been the broad objectives of political leadership for the state bureaucracy during the 1920s and, within those objectives, the roles in which the lower classes had been cast. I next evaluate the degree to which those objectives were realized, then discuss the implications of the success or failure of the political leadership to assert its control over bureaucratic operation and staffing. Finally, I identify what seem to have been the most important consequences of the actual patterns of bureaucratic change so far as upward social mobility in Russia was concerned.

Objectives for Administrative Development

We may characterize the Soviet political leadership's collective objectives for bureaucratic reorganization under four heads: (1) debureaucratization, (2) structural modification, (3) proletarianization, and (4) political control (Communization).

It may seem anomalous to begin a discussion of bureaucratic development with a discussion of *debureaucratization*. Marx and Engels dealt extensively with this objective, easily recognized in the Marxian context as part of the millennial withering-away of the state. As noted earlier, Lenin too addressed debureaucratization, as did Bukharin and Trotsky.

Soviet political leaders during the 1920s seem to have used the term "bureaucratic" only in the pejorative senses implying self-interest, insensitiveness, and organization-for-its-own-sake, in spite of the role the term was already playing in serious social science research into contemporary political and organizational behavior. The tendency seems to have been to apply "bureaucracy" and "bureaucratic" to situations where the actions of individual officials—whether of the Party or of the state—were questionable, clumsy, or, especially, exclusive and cliquish. For example, during the spring and summer of 1923, when Trotsky, Zinoviev, Stalin, and their various allies were maneuvering for control, "bureaucratic" and "bureaucratism" were epithets that stood for the cliques and cabals of the competition.[4] Later, of course, the charge of bureaucratization of the Party, made by Trotsky against Stalin, would form the leitmotif of a vocal, largely emigré or foreign opposition.[5]

Alternatively, "bureaucratic" was used to describe the tendency of postrevolutionary society to retain fragments of prerevolutionary social and political values. For example, Lenin identified the Old Regime with bureaucracy: "The Bureaucracy has been smashed. The exploiters have been pushed to one side. But the cultural level is not rising, and therefore the bureaucrats occupy their old places."[6]

Debureaucratization as a political objective thus tended to focus on the actions of individual officials and informal groups of officials, and especially on the values that one might attribute to those actions. As this study is examining larger elements of bureaucratic behavior and structure than the action of individuals, I am not specifically concerned with debureaucratization in the senses noted here. The last chapter of this book does however, raise the question whether debureaucratization is a feasible objective for postrevolutionary Soviet society.

Structural modification. Bureaucracy is not the only possible type of administrative structure, as we saw in Chapter 1. Although bureaucracy—a hierarchical distribution of specialized functions—is probably the most usual in large organizations, other common patterns are the familial and collegial structures. As Jerry Hough shows explicitly and Marc Raeff implies, moreover, in the context of Russo-Soviet administration the term "bureaucracy" invokes several different structural models.[7]

Soviet political leaders—not to mention the leadership of the Provisional Government—seem to have expected the postrevolutionary administration of finances, education, health care, scientific research, and so forth, to take some form other than the bureaucratic one. On many occasions Lenin discussed structural alternatives in general terms.[8] And as preceding chapters have shown, there were substantial structural changes in civil administration beginning in 1917: functional specialization and appropriate reorganization of personnel resulted in

---

4. See, e.g., Isaac Deutscher, The Prophet Unarmed (New York, 1965), 110–18; and T. H. Rigby, Lenin's Government (Cambridge, 1979), 11.
5. Leon Trotsky, The Revolution Betrayed (New York, 1937), 89–102.
6. Sol'skii, "NOT i voprosy deloproizvodstva," 47.
7. Jerry F. Hough, The Soviet Union and Social Science Theory (Cambridge, Mass., 1977), 49–78; Marc Raeff, "Bureaucratic Phenomena of Imperial Russia," American Historical Review 84, no. 2 (1979), 399–411.
8. V. G. Afanas'ev, "V. I. Lenin o nauchnom upravlenii obshchestvom," Voprosy filosofii, no. 1 (1974), 17–29; Sol'skii, "NOT i voprosy deloproizvodstva."

an administrative apparatus that looked quite different from the one that was in place in 1916; moreover, the roles of professional specialists changed significantly and permanently. But as Rigby has noted, when Lenin addressed himself to the execution of specific tasks, he seemed to think in terms of modified versions of common, ready-to-hand organizations—that is, bureaucracies.[9] The substantial growth of hierarchical organizations in Soviet state administration during the 1920s suggests that organizational techniques and the sheer practical demands of completing certain tasks—rather than theory—crowded out the structural alternatives.[10]

Another aspect of the problem of what kind of organization was appropriate to the postrevolutionary era is both more involved and possibly more fruitful for further investigation. We have seen substantial evidence that control of administrative activities by technical or professional specialists grew and intensified during the 1920s, even by comparison with the second half of the nineteenth century in the tsarist bureaucracy. Given Lenin's emphasis on the need for "scientific" administration of society, and the interest of Nadezhda Krupskaia (while she was politically active) and others in the technically oriented administrative theories of F. W. Taylor, one must at least question whether the further extension of technical control may have been partly a product of the objectives of political leaders. Once more, however, this problem is not of direct concern here, since resolution of the question has to depend upon analysis of the roles of departmental, vice-commissarial, and commissarial administrators throughout the 1920s.

The objective of *proletarianization* speaks to the very heart of the revolution and of our interest in it. In spite of the evident critical importance of proletarian power to the success of a Marxist revolution, Rigby shows that proletarianization as a goal—at least at the level of the Council of People's Commissars—receded during Lenin's tenure as chairman.[11] Fitzpatrick maintains that large-scale proletarianization was a phenomenon associated with the era following Stalin's consolidation in power—that is, after 1928.[12] She argues, moreover, that the Red managers who were in positions of administrative authority over the well-educated bourgeois specialists before the famous Shakhty Trial

---

9. T. H. Rigby, "The Birth of the Central Soviet Bureaucracy," *Politics: Australasian Political Studies Association Journal* 7, no. 2 (1972), 123–35.

10. Stephen Sternheimer, "Administration for Development," in W. M. Pintner and D. K. Rowney, eds., *Russian Officialdom* (Chapel Hill, N. C., 1980), 316–54; Pintner and Rowney, "Officialdom and Bureaucratization: Conclusion," in ibid., 369–80.

11. Rigby, *Lenin's Government*, chap. 11.

12. Sheila Fitzpatrick, "Stalin and the Making of a New Elite, 1928–1939," *Slavic Review* 38, no. 3 (1979), 377–402, and *Education and Social Mobility in the Soviet Union* (Cambridge, 1979), chaps. 7–9.

(1928) were saddled with serious disabilities. Either they hailed from questionable (nonproletarian) backgrounds themselves, or they could be viewed as ignoramuses, helpless pawns in the hands of the bourgeoisie.[13] The question of the proletarianization of the administrative apparatus is thus of some importance in any effort to characterize bureaucratic change and is therefore discussed below at length.

Political control (Communization). Like proletarianization, the domination of the civil administration by the political leadership, principally by the Communist Party, could be taken as a given of the revolution. It is assumed here that both the Party leadership and its rank and file were interested in some measure of control over state administration, at least to the degree that politicians in other states are interested in controlling civil administration.

Possibly it can be agreed that political interest in control over the administrative apparatus was twofold: first, control over operations, particularly those that bore upon segments of the society which merited political concern, such as the rural poor and the urban proletariat; second, control over personnel, especially over appointments to administrative offices. These objectives are not always completely separable. Control over personnel is often, though not always, sought as a means of controlling operations more generally. As we have already seen, there is some evidence that the transfer of prerevolutionary officials to positions in the Soviet administration was governed not only by circumstances of organizational novelty but by political considerations as well. New organizations and politically sensitive agencies were unlikely to serve as homes for tsarist holdovers, probably because of considerations of political control. In addition, however, control over appointments may also be sought for the much more immediate objective of establishing political patronage. An aspect of this relationship between the Communist Party and the state administration during the 1920s is also of direct concern here.

Social Classes and Their Relative Status

Since part of the problem is to determine whether there was any administrative impact on class relations and upward mobility, it is important to be explicit about what is meant by the specific class categories worker, peasant, and white-collar worker. The problem of definition includes not only the dictionary meaning of the three words but also their relative meaning, or status with respect to one another.

Finding the relative meaning is a surprisingly straightforward matter

---

13. Fitzpatrick, "Stalin," 379–80.

of combining occupation data from the 1920s with social class data—
or what Rigby identifies as the employment category that an individual
held before joining the Party[14]—in order to indicate social mobility:
that is, investigating whether and to what extent change in social status
occurred by comparing the "current" employment patterns of a given
group with its "social situation," or status-of-origin data. During the
1920s the Party diligently sought information on this question of the
social origin and previous employment history of state functionaries,
Party members, and their families; accordingly, it commonly obliged
members to document their social class background (employment of
parents, for example) and their own employment background prior to
their joining the Party.[15] Because party surveys often presented these
data together with current employment data for Party members and for
state functionaries, we can compare the two bodies of information for
indications and measures of both occupational and social mobility.

First, let us try to establish both the meaning of the three broad job
categories worker, peasant, and white-collar worker and an order of
status, or preference, for them. Briefly, "workers" included most categories of urban skilled and unskilled labor and some categories of
rural day labor *(batrak)*.[16] "Peasants" comprised three socioeconomic
levels of workers on agricultural land: poor *(bedniak)*, middling *(sred-
niak)*, and well-to-do *(kulak)*. These categories are generally distinguished from agricultural day labor because peasants owned land and
implements or were members of rural communes that owned land and
implements.[17] "White-collar worker" *(sluzhashchii)* tended to include
anyone whose social background identified him or her neither as a
member of the former exploiting (bourgeois or noble) classes nor as a
worker or peasant. In practice, the survey data commonly apply the
term to a broad range of administrative, intellectual, and white-collar
employees of the state or of non-state institutions such as cooperatives.
For this reason one often finds the term *sluzhashchii* translated not
literally—"servitor"—but functionally: "white collar worker," "administrator," "official," or simply "employee."

---

14. T. H. Rigby, *Communist Party Membership in the USSR, 1917–1967* (Princeton, N. J., 1968), 160–61.

15. V. Z. Drobizhev and E. I. Pivovar, "Large Data Sources and Methods of their Analysis," in D. K. Rowney, ed., *Soviet Quantitative History* (Los Angeles, 1983), 147–68.

16. Frank Lorimer, *Population of the Soviet Union: History and Prospects* (Geneva, 1946), Table 37, App. A.

17. D. J. Male, "The Village Community in the USSR, 1925–1930," *Soviet Studies* 14, no. 3 (1963), 225–8; Ivan Koval'chenko and N. Selunskaia, "Large Data Files and Quantitative Methods in the Study of Agrarian History," in Rowney, *Soviet Quantitative History*, 47–74; Lorimer, *Population of the Soviet Union*, Table 37, App. A.

From the point of view of Communist social theory, Soviet workers occupied the highest notch on the scale of social status. Soviet society, the product of a Marxist revolution, sought to enhance—to glorify, really—the status of industrial workers in many different ways. Nevertheless, it is evident that then as now the concept of the worker as a social elite met with considerable skepticism. For example, the tendency of white-collar workers and executives to be accorded high status by others, and to accept or seek out high status, was noted—disapprovingly—in Party congresses and publications during the 1920s.

As Alex Inkeles shows, during the 1930s and 1940s traditional marks of status difference were reintroduced into Soviet society.[18] But tangible differences existed even before that. For example, there were significant salary differences among low-and high-status administrative personnel throughout the 1920s, as Sternheimer notes.[19] more important, there were substantial differences between aggregated mean administrative salaries and mean industrial (workers') wages throughout the period. Even local and district administrative officials (of soviet *ispolkomy*) drew salaries that were typically higher than mean worker wages, according to data tabulated by Sternheimer.[20]

In all, the implication that manual labor was worth less to society than relatively intellectual or nonmanual employment seems inescapable, just as the range of differences placed a considerable strain on the notion of the desirability of working-class labor as distinct from intellectual or managerial work. Thus, the three major categories of workers, peasants, and white-collar workers are taken here to indicate three distinct levels in society and in labor, with workers holding a middle position between the other two classes and peasants occupying the lowest rung on the status ladder. Although very highly skilled workers (toolmakers, for example) could earn as much or more than a lot of administrators or white-collar employees (especially those in local and district administration), it seems safe to say that the majority of workers did not. We thus conclude that the members of the working class who found white-collar employment may be seen as having advanced themselves both occupationally and socially.

Class, Party, and Official Employment in the 1920s

Party membership, in broad terms, was very much a two-way street so far as influence over officeholding was concerned. That is, Party

18. Alex Inkeles, *Social Change in Soviet Russia* (Cambridge, Mass., 1968), pt. 3, "Social Stratification."
19. Sternheimer, "Administration for Development," 333–34.
20. Ibid., Table XII.5.

TABLE 19.
Communists among officials, 1922 and 1929

	Moscow 1922	All Union 1929	Republic 1929
Top grades	13%	48%	49%
Specialists	5	19	17
Junior grades	4	28	22

Sources: V. I. Vasiaev et al., *Dannye perepisi sluzhashchikh 1922 g.* (Moscow, 1972); Ia. Bineman and S. Kheinman, *Kadry gosudarstvennogo i kooperativnogo apparata SSSR* (Moscow, 1930).

NOTE: with the creation of the Union of Soviet Socialist Republics, a new tier was added to the traditional tsarist three-level administrative structure. This new "republican" administration employed officials both in Moscow and in the capitals of the "Union Republics" (Belorussian, Kazakh, Kirgiz, Russian, Transcaucasian, Turkmen, Ukrainian, and Uzbek). "All Union" officials in Bineman and Kheinman's *Kadry* are taken in this book to mean those who worked in Moscow for the central offices of the commissariats. In some instances these include "Officials of the Union Republics" who appear to have been stationed in Moscow but who were employed by republic commissariats. "Republic" officials in Bineman and Kheinman are taken to be those who were working in republic capitals for commissariats of individual republics. *Oblast, Okrug,* and *Raion* officials in Bineman and Kheinman are taken to be officials who worked for commissariats in either regional or local offices.

members qualified to fill positions in the state administration could be given preference over non-Party persons in appointments. At the same, time officeholders or those intending to seek appointments to administrative positions in the 1920s could attempt to gain entrance into the Party in order to enhance their career prospects. This phenomenon is variously reflected in the characteristics of the personnel who held the offices, but as a rule of thumb, we could say that Party membership (*partiinost'*) reflected the sensitivity of a given person, a position, or an organization to political issues. High organizational *partiinost'*—that is, an office in which a high proportion of the staff belonged to the Communist Party—could be taken to reflect relatively great political sensitivity; low *partiinost'*, the opposite. Still, there were important exceptions to this rule.

Much of the following discussion is based on a comparison of two somewhat different profiles of officials. The first, the compilation by Vasiaev and others of 1922 data is based on a relatively small sample geographically restricted to the Moscow area.[21] The second is the last comprehensive study of Soviet officialdom of the 1920s, completed by Ia. Bineman and S. Kheinman in 1929.[22] In spite of the difficulty of comparing these studies and interpreting them jointly, the findings—especially when tempered by findings from additional sources—make the effort worthwhile.

Table 19 offers a direct comparison of affiliation with the Communist

21. V. I. Vasiaev et al., *Dannye perepisi sluzhashchikh 1922 g.* (Moscow, 1972).
22. Ia. Bineman and S. Kheinman, *Kadry gosudarstvennogo i kooperativnogo apparata SSSR* (Moscow, 1930).

Party, level by level, for all "central" commissariats in 1922 and again as of October 1929. Party members fell into two categories, full members and candidate members. While the respective ratios of candidate (or Communist Youth) members to regular members reflect variation in the age of given groups, the increase in proportion of Party members at each of three levels of administration is enormous.

Another pattern emerging from these data is the greater association of Party membership with elite status than with simple participation in the administrative apparatus; a much higher proportion of Bolsheviks appears among top officials than in other categories. According to additional data from Bineman and Kheinman, this pattern intensified throughout all the territorial levels of administration. Thus about half of all *oblast* (provincial) officials and about 70 percent of all top *okrug* and *raion* (district and township) officials were either Bolsheviks or Communist Youth (*Komsomols*). Comparable figures for junior-grade officials were on the much reduced order of 20–25 percent.

The data on Party affiliation highlight still another important pattern: generally, at all levels of administration, individuals in specialist and scientific roles were less likely to have been Party members or candidates. The data in Table 19 indicating more than 80 percent non-Party officeholders among the group in 1929 are replicated at the *oblast* and subprovincial levels. Nowhere did Party membership exceed 20 percent; typically, it hovered at about 10 percent. While some of this manifest non-Party tendency is possibly due to either the junior status or junior age of these officials (technically, acceptance into the Party required a period of trial membership), most of it must surely have been the result either of their failure to "qualify" from the point of view of class or of their relative independence from political pressure. Well educated, they may have been members of the *ancien* sub-elite; technically educated, they may have been indispensable to the Bolshevik administration.

Class or family background data for the 1929 cohort are also of great interest. Top officials in 1922 who admitted to previous experience in the state administration—and who therefore could not be classified as workers or peasants—were, at a maximum, 68 percent of the total, with 29 percent admitting to such a background before 1917. The comparable category for 1929 is 71 percent of the total, including "children" of servitors. The children's category, by itself, could have made up the difference between the 1922 and the 1929 populations. But even more striking is the comparison between 1922 and 1929 of specialists' and juniors' background in state service. In 1922 the maximal figures were 36.4 percent (specialists) and 27.1 percent (juniors); in 1929, the comparable All-Union and

TABLE 20.
Age distribution of commissariat officials, Moscow, 1922

	N	Under 26	26–30	31–35	35–50	Over 50
Top grades	3,064	11%	19%	20%	39%	10%
Specialists	2,037	19	18	15	32	12
Junior grades	1,833	27	18	16	27	12

Source: V. I. Vasiaev et al., *Dannye perepisi sluzhashchikh 1922 g.* (Moscow, 1972).

TABLE 21.
Age distribution of officials in All Union commissariats and central administration, 1929

	N	Under 25	25–29	30–34	35–44	Over 44
Top grades	2,348	3%	9%	18%	42%	28%
Specialists	4,895	5	18	21	33	23
Junior grades	3,585	13	29	21	23	14

Source: Ia. Bineman and S. Kheinman, *Kadry gosudarstvennogo i kooperativnogo apparata SSSR* (Moscow, 1930).

Union Republic figures were 76 percent and 70 percent (specialists) and 58 percent and 57 percent (juniors)—about twice as high, on the whole, as the level indicated by the 1922 data. Central administration in the USSR during the course of the 1920s would clearly appear to have fallen into the hands of a population whose members considered themselves—or were classified—as state servitors in something more than a mere employment sense.

A look at the relevant age data suggests additional conclusions. Tables 20 and 21 describe a much more mature population of top-grade and specialist officials in 1929 than in 1922. Were it not for the high levels of turnover already discovered for the period, one would be tempted to suggest that this is the 1922 cohort merely grown older in office. And indeed, these officials may be the 1922 cohort in part, since turnover would include movement from one office to another even if the second office were still in the bureaucracy. In other words, what one sees in Tables 20 and 21 may be partly accounted for by movement from one agency to another rather than by a definitive move out of the bureaucracy. But the data clearly show more than that. For if these older cadres were also predominantly state servitors in a social sense, then those social origins must have been prerevolutionary. In other words, whatever change occurred in the central top-grade and specialist groups

TABLE 22.
Level of education attained by commissariat officials, Moscow, 1922

	N	Higher	Secondary	Primary	Other/unknown
Top grades	3,015	57%	26%	9%	8%
Specialists	1,907	61	22	13	4
Junior grades	1,673	41	29	25	5

Source: V. I. Vasiaev et al., *Dannye perepisi sluzhashchikh 1922 g.* (Moscow, 1972).

during the 1920s was accounted for by an influx—a re-influx?—of ancien régime officials similar to that of 1922 but more substantial.

It is important to note the sharp contrast in ages between the junior-grade officeholders in 1929 with the comparable group in 1922: about 61 percent were under thirty-five in 1922; slightly more (63 percent) were under thirty-five in 1929, seven years later. Thus, if a proportion of the top officials and specialist were graying in office, the juniors were certainly not. If anything, they were growing relatively younger; the generational distance between them and their senior-level colleagues was increasing.

The educational data (Tables 22–24) suggest both professional and social comparisons between the 1922 and 1929 cohorts. It is clear, for example, that both top-grade officeholders and specialists in 1929 were slightly—but probably only slightly—less highly educated than their counterparts in the Russian Republic in 1922.

The massive difference in educational backgrounds emerges from comparison of the Junior cohorts of 1922 with those of 1929. There the decline in higher education has been precipitate: about 13 percent had university training in 1929 All-Union commissariats as against 40.7 percent at the comparable job level in 1922. These data again underscore the emerging differences between the top and specialist grades on the one hand, and the juniors on the other. The fact that the juniors had not only become much younger, relatively, but much less well educated points to both a generation difference and possibly a class difference between these categories. The 1929 top and specialist cohorts, while distinct with respect to their membership in the Communist Party, together formed a sharp contrast with their junior colleagues. The top and specialist officials were not only older and better educated but, evidently, much more likely to have received their education under the last tsar. Indeed, a conservative estimate would be that at least a third of them received all of that education before 1917. Given the upper-class bias of educational opportunities before 1917, this conclu-

TABLE 23.
Level of education attained by commissariat officials, All Union, 1929

	N	Higher (−1918)	Higher (1917−)	Some secondary	Some primary	Other/unknown
Top grades	2,348	23%	36%	22%	16%	3%
Specialists	4,895	22	40	28	8	2
Junior grades	3,585	2	11	46	34	7

Source: Ia. Bineman and S. Kheinman, Kadry gosudarstvennogo i kooperativnogo apparata SSSR (Moscow, 1930).

TABLE 24.
Level of education attained by commissariat officials, Republic, 1929

	N	Higher (–1918)	Higher (1917–)	Some secondary	Some primary	Other/unknown
Top grades	3,399	16%	32%	26%	20%	6%
Specialists	5,435	15	34	39	8	3
Junior grades	4,642	1	12	47	28	12

Source: Ia. Bineman and S. Kheinman, Kadry gosudarstvennogo i kooperativnogo apparata SSSR (Moscow, 1930).

sion strengthens the suspicion that the top and specialist officials came from privileged class backgrounds.

The holders of junior offices were not merely younger and less highly educated but evidently crippled in an organizational sense, in spite of the fact that they were more likely to have received their education under Communism. By comparison with the standards of educational achievement set by their organizational superiors or by their junior-level predecessors in 1922, the juniors of 1929 were in no sense fit to compete for the positions tied down by their aging tsarist-trained bosses. What seem to have been emerging between 1922 and 1929, in other words, were not merely differences of a numerical order or even differences in the order of organizational experience usually conferred by seniority. What was apparently emerging was a de facto class difference between the two groups. For if by class distinctions one understands not only present socio-economic status but life opportunities, opportunities for the future, then class distinctions were clearly emerging in the central administrative apparatuses in the 1920s.

It is true that status distinctions are always present in hierarchical organizations, but the way in which they are established can be very important. One can say that in the tsarist system the most significant organizational differences in status came about because of seniority and education—always keeping in mind that the social elite had access to the best education. On the average, those who had obtained the proper educational credentials could expect to achieve superior organizational status as their seniority increased.

My suggestion is that, to anyone who was watching closely in the 1920s (and we may presume that the junior-level personnel were watching very closely), career expectations based on the idea of improving status with increasing seniority were threatened both by the educational gulf between the top and junior categories and by the top administrators' increasing seniority, even in the face of a constant turnover of individuals. Given that the juniors did not possess educational qualifications remotely comparable to those of their seniors, and given that the seniors were evidently hanging on to their status, if not to specific offices, the organizational future could not have looked bright to the vast majority of juniors as the 1920s wore on. Given, moreover, that the top officials and the specialists were more likely to be products of the *ancien régime* than were the juniors and that the specialists even lacked the credentials of good Communists (Party membership, remember, was higher among many juniors than among specialists), it is entirely conceivable that the growing class distinction between the top and Specialist groups on the one hand and the juniors on the other was also tinged with class hostility.

The Odor of an Old Elite and the Ambition of a Young Pre-Elite

Students of the revolutionary era have already written about class hostility both in Russia and in the young Soviet republic, and it seems unnecessary to belabor the matter here. What I wish to suggest is that owing to its comparatively formal and inherently hierarchical structure, the state administration of the 1920s forged in spite of itself an environment in which class differences could be very sharply articulated and personally experienced. The career experiences of thousands of state officials bore more than a trace of those social status differences that could be expected to evoke class hatred of "holdovers"—less, perhaps, as individuals than as embodiments of the social characteristics, class values, and "bureaucratic" behavior associated with long-remembered and deeply cherished hatred for the old elites, characteristics that even upper-class Old Bolsheviks could not entirely shed in the eyes of peasants and workers.

One of the standard interpretations of the transition from Leninism to Stalinism in the 1920s is that Stalin's advantage over his competitors for Lenin's mantle consisted in his control over local Party cadres and, through them, over the Party Congresses and the Central Committee.[23] The basis of this control is said to have been Stalin's influence over the Party Secretariat with its power of appointment to local Party offices and, through the *nomenklatura*, the power to establish procedures for appointment to non-Party offices. Rigby has recently reinforced this interpretation, noting that a political fabric into which cliques wove themselves was a fact of early postrevolutionary provincial and local party life. Stalin was ultimately victorious, Rigby says, because he was the man at the center who had the wit to take advantage of "the transformation of provincial cliques into clientist followings under the patronage of Moscow-chosen party secretaries."[24]

The experience of a successful combination of Party affiliation and executive administrative positions for an ex-peasant or an ex-worker was certainly achieved when Stalin's influence on the formation of the Party apparatus was building. I am suggesting that the data already

23. T. H. Rigby, "Early Provincial Cliques and the Rise of Stalin," *Soviet Studies* 33, no. 1 (1981), 3–28; Leonard Schapiro, *The Communist Party of the Soviet Union*, 2d ed. (New York, 1970); Adam B. Ulam, *Stalin: The Man and His Era* (New York, 1973).
24. Rigby, "Early Provincial Cliques," 25.

examined describe a substantial population of officials who knew from direct personal experience the rewards associated with Party membership and who may well have been prepared to exploit this knowledge if fresh opportunities should arise. The events of succeeding years—collectivization, the drive for lower-class education of the late 1920s, the purges of Party leadership—may all be seen as opportunities for a young, ambitious, frustrated population to exploit the connection between membership in "Stalin's Party" and upward career mobility.

Using the data examined above, one can identify the would-be pre-elites of the civil bureaucracy in the 1920s, those who would succeed to administrative authority in the 1930s if the way should, by some chance, happen to open. In speaking of them, it is important to recognize that one may well be speaking of one of Stalin's most powerful political assets. Indeed, one may even be speaking of a group whose support Stalin was perceptive enough to seek out through his policies in the 1920s. I will return to this point, but it is useful to note here that it is not necessary to imagine that the pre-elites of the 1920s were a social or political category that Stalin himself created or consciously manipulated, as Seweryn Bialer appears to suggest and as Trotsky asserted again and again.[25] To be sure, the policies of the Central Committee in the 1920s encouraged the formation of Party cadres from among young workers, who would then have acquired sufficient clout to seek administrative jobs. This would seem to be an ordinary and predictable part of revolutionary politics: one expects that a revolutionary party will enhance its power by acting in the interests of the social classes who are its main support. While it is not excluded that these were policies devised by an extraordinarily prescient and manipulative Stalin, they could just as well have been formulated by broader sectors within the Party for the Party's own well-being. Alternatively, they might simply reflect the response to motivations that affected the decisions of tens of thousands without any single individual's conscious planning.

It seems plausible that the existence of a young, socially distinct, ambitious underclass within the bureaucracy actually changed what is assumed to be the normal direction of the flow of power in the Stalin days. Stalin and other members of the Soviet political elite may actually have felt pressure from this underclass, pressure that galvanized the Party leadership into taking steps to secure rapid social change that it might not otherwise have taken. In any case, the existence of a broad, well-placed spectrum of potential support for radical action on highly

---

25. Seweryn Bialer, *Stalin's Successors: Leadership, Stability, and Change in the Soviet Union* (New York, 1980), pt. 1, "The Mature Stalinist System," 9–27.

controversial issues such as education, agriculture, and industrialization policies would go far to explain the haste with which certain programs were executed at the end of the first decade of Communist rule. In short order, in the late 1920s, radical programs for the education of the lower classes, comprehensively controlled industrial development, and the collectivization of agriculture were introduced and implemented with great dispatch.

To address these issues directly, let us now bring the role played by Party membership to the fore. Who were the Communists of the 1920s and what was their relation to state administration? If the Communist Party was supposed to be a party by and for the lower classes, did this aspiration translate into improved status among the lower classes? Do the data show that the lower classes were doing anything but picking at the nether fringes of the postrevolutionary power apparatus? Were workers and peasants actually moving out of the social categories into which they were born in Russia? Were they moving upward in significant numbers? If we enlarge the scope of our attention to include not only the civil bureaucracy but the postrevolutionary state's principal political organizations, the All-Russian Communist Party and the Communist Union of Youth (Komsomol), an answer to this question—really a fundamental question about the success of the revolution—emerges.

The Class Composition of the Political Leadership

Before analyzing the patterns of transfer between social class of origin and type of employment for the Party and the civil administration as a whole, one must ask whether the characteristic of transfer out of the working and peasant classes was apparent at the most elite levels of state and Party administration. Were the Central Committee of the Communist Party and the Council of People's Commissars open to an influx of lower classes after the revolution? Or, did they, like the new commissariats, tend to have significant proportions of holdovers, intellectuals, and former sub-elites? As it happens, quite a lot of social mobility was evident there, and its timing is of considerable interest.

As we know from Rigby's work on the early Soviet government, there was a tendency within the highest, most political levels of the government to move away from lower-class representation during Lenin's era (through 1922).[26] The number of former workers or peasants who had successfully managed the transition to either the elite administrative or political levels declined following the revolution or, at best, stayed

---

26. Rigby, *Lenin's Government*, 150, Table 5.

low. Leadership positions in the Party and the state apparatus were in the hands of the comparatively well-educated middle and upper classes.

Certainly Lenin and many of the prerevolutionary Bolshevik leaders were themselves rather far removed, both culturally and socially, from the proletariat and the peasantry. Thus, it may not really be surprising that Rigby found the Lenin government growing progressively less and less working-class up to the moment when Lenin was incapacitated by a cerebral hemorrhage. These people were likely, one would suppose, to have surrounded themselves with and entrusted authority to others whose life experiences, social behavior, and intellectual aspirations were similar to their own—provided, of course, that there was the necessary congruence of political values.

But what about the middle and later 1920s? Did some of those who had left worker or peasant status actually make their way to elite status in the government or in the Party after Lenin's departure from the scene? In fact, Trotsky complained specifically about the "green and callow" flood of new recruits who emerged at the moment of Lenin's death to seize positions in Soviet officialdom.[27] An influx of even a few members of the lower classes into the seats of power and influence might well have looked green and callow to an upper class intellectual like Trotsky, who was known to suffer fools badly and whose relatively elevated social status permitted him to define fools broadly. But does the evidence indicate increased, stable, or decreased upward mobility into the governing Party and civil elites of the Soviet Union in the middle 1920s?

In 1922 three-quarters of the membership of the Council of People's Commissars (SOVNARKOM), according to Rigby, was of upper-class origin.[28] Combining biographical data from several sources makes it possible to characterize the class or employment origins of the council for much of the remainder of the 1920s.[29] By 1923 the proportion of members of an expanded Russian Republic Council who were of white-collar, bourgeois, or noble origin had dropped slightly, to 68 percent. By 1927 the proportion of *ancien* upper classes in the now combined USSR and Russian Republic Council had dropped much more, to less than half. Correspondingly, the proportion of members of working-class or peasant background had risen to 55 percent. This figure for lower-class participation is slightly higher than for the government as a whole; it is much higher than for the central administrations of the state bureaucracy (see Table 31 below). Moreover, almost half the total of

---

27. L. D. Trotsky, *Stalin: An Apraisal of the Man and His Influence*, ed. and trans. Charles Malamuth (New York, 1967).

28. Rigby, *Lenin's Government*, chap. 11.

29. *Vsia Moskva: Adresnaia i spravochnaia kniga na—god* (Moscow, 1923–27); *Who Was Who in the USSR* (Methuen, N. J., 1972).

twenty-four council members had managed to hang on in the SOVNARKOM from 1923, and they are a sufficiently large proportion to account for most of the increase in mean age of the council from about forty-three to forty-five years. Of the new members, however, the clear majority were relatively young workers or peasants by background.

We find the same pattern within the principal agency of the central Party organization, the Central Committee. The class background of thirty-four candidates and members in 1923 can be satisfactorily identified by combining records from the list of Central committee members at the twelfth and fifteen Party Congresses.[30] Of the 1923 cohort, about twenty (roughly 60 percent) were born to fathers who were peasants, workers, or skilled workers. By 1927 the size of the organization had increased substantially, from fifty-seven to eighty-nine members and candidates. So too had the proportion of members and candidates who were workers or peasants by background: in 1927 the figure stood at about 70 percent.

It is quite possible that these events were manipulated by Stalin as a prelude to moves in the direction of the proletarianization of the state administration. But given that Stalin was engaged in a major struggle for control of the Central Committee and the government during much of this time, it seems at least questionable that even his position as Party First Secretary would have given him so much authority over the highest levels of *nomenklatura*—the Central Committee and the Council of People's Commissars—so early. In any case, it is important to note that the class reorientation of these organizations is more directly associated, from the viewpoint of timing, with Lenin's departure than it is with the introduction of policies directed toward worker promotion at the end of the 1920s. This is so despite the impression one derives from Rigby that proletarianization beginning with the "Lenin Enrollment" in 1924–25, was not a success.[31]

Senior Officials: Two Administrative Elites

A comparison of some characteristics of senior officeholders in the Party and civil administration during the 1920s may help clarify the political dynamics that were operating during the middle and later years of the decade—in particular, the degree, timing, and location of administrative proletarianization as well as the extent and net effect of political control over administration.

---

30. *Bol'shaia sovetskaia entsiklopediia*, 1st ed. (Moscow, 1926–), 9: 554–57; *Who Was Who in the USSR*.
31. Rigby, *Communist Party Membership*, chaps. 3, 4.

TABLE 25.
Level of education attained by All Union and Republic officials, 1927 and 1929

	N	Some higher	Some secondary	Some primary
Party members[a]	3,683	15%	29%	56%
Top civil ranks[b]	5,747	53	23	21
TsIK members[c]	209	59	20	21

Sources: Central Committee, All-Union Communist Party (bolsheviks), Statistical Division, Sotsial'nyi i natsional'nyi sostav VKP(b) (Moscow, 1928); E. Smitten, Sostav vsesoiuznoi kommunisticheskoi partii (bol'shevikov) (Moscow, 1927); Ia. Bineman and S. Kheinman, Kadry gosudarstvennogo i kooperativnogo apparata SSSR (Moscow, 1930).
[a]Data for 1927.
[b]Data for 1929; "unknown" of 2.8% not included.
[c]Soviet central executive committees, data for 1929.

The data presented here are drawn from several different sources, but the most important are of two kinds: first, the Party censuses taken during the 1920s, especially the major census of 1927;[32] second, the large study prepared by Bineman and Kheinman and published in 1930.[33] Extremely useful as a source of data on the characteristics of civil administration generally, especially the provincial and local administrations during the later 1920s, this second source is somewhat more problematical for the purposes of my study than the Party censuses. This is because the timing of the movement of lower-class social groups into offices of the state administration—specifically, whether it occurred before or after the inauguration of the worker promotion program in 1928—is important to the analysis. This detail is of sufficient importance to warrant a brief digression.

The administrative census upon which the Bineman and Kheinman study is based dates from the fourth quarter of 1929. Comparison of these data with earlier studies is difficult and somewhat risky. For example, the study by Vesiaev's group, used earlier, is based upon a relatively small sample (whose characteristics as a sample are discussed by the Soviet authors at length) drawn in 1922.[34] The numbers of officeholders counted in the Bineman and Kheinman study are very large, by contrast, and they include the entire apparatus—central, provincial, and local. In spite of its obvious problems the Bineman and Kheinman study seems suited to my purposes. The pattern of lower-level officeholding, in particular, is so distinct from that of the pre-1917 administration that it is unreasonable to think all these changes could have

32. Central Committee, All-Union Communist Party (bolsheviks), Statistical Division, Sotsial'nyi i natsional'nyi sostav VKP (b): Itogi vsesoiuznoi partiinoi perepisi 1927 goda (Moscow, 1928); E. Smitten, Sostav vsesoiuznoi kommunisticheskoi partii (bol'shevikov): Po materialam partiinoi perepisi, 1927 goda (Moscow, 1927).
33. Bineman and Kheinman, Kadry, 86ff.
34. Vasiaev et al., Dannye perepisi, 3–49.

TABLE 26.
Level of education attained by officials in regional administrations, 1927 and 1929

	N	Some higher	Some secondary	Some primary
Party members[a]	8,111	3%	17%	80%
Top civil ranks[b]	27,783	26	26	44
TsIK members[c]	2,360	13	19	68

Sources: See Table 25.
[a]Data for 1927.
[b]Data for 1929; "unknown" of 2.8% not included.
[c]Soviet central executive committees.

TABLE 27.
Level of education attained by officials in local administrations, 1927 and 1929

	N	Some higher	Some secondary	Some primary
Party members[a]	11,275	0%	6%	94%
Top civil ranks[b]	17,129	3	23	51
TsIK members[c]	9,123	2	16	82

Source: See Table 25.
[a]Data for 1927, raions and volosts.
[b]Data for 1929, raions; "unknown" of 2.8% not included.
[c]Soviet central executive committees, raions.

occurred in a matter of twelve to fifteen months. Therefore, I propose that the data reflect the social characteristics of the civil administration at least from the middle 1920s and that they are chronologically comparable to the Party census of 1927.

Tables 25–27 show the pattern of shifts in the educational background of Party members who held active, functional positions in senior state service, according to the Party census of 1927. The pattern may be compared with the educational characteristics of senior personnel in state administration and of members of local soviet executive committees, known as *ispolkomy* (for *Ispolnitel'nye komitety*) or TsIKs (for *Tsentral'nye ispolnitel'nye komitety*).

One could summarize the contrasts by saying, first, that educational levels were generally much lower for Party members and candidates in state service than for the service as a whole; and second, that the distance of all officials, but especially of Party officials, from the center was generally inversely related to educational level. That is, the civil servants generally and the Party members in particular who were employed in lower and more provincial offices were relatively uneducated. Wide variation in qualifications was, possibly, as common in large segments of the Soviet administrative system as it had been under the tsars—but the degree of variation was much greater. Overall, the educational levels of the later 1920s were comparable not to those of 1890 or the turn of the century but to those of the mid-nineteenth century, as indicated by Pintner's data for the earlier era.[35]

These data indicate a major change in the composition of staff. One way of understanding what happened is to imagine that the first generation of Russian industrialization, the generation that produced the cadres who successfully made the transition described in the preceding chapter, had simply not existed. There would have been a senior administration, of course, but it would not have had the same education and experience as the generation of 1880–1914 because well-educated staff would have been too rare.

But the effect of the revolution on the senior administration was more comprehensive than a mere restructuring of the distribution of educational characteristics would imply. It was not only the technical and professional upper-class elites who were displaced; removals went right across the board, and they were made on the basis both of class and of previous occupation. It was as though, at some bureaucrats' seaside convention, the entire senior cohort of 1917, ranks 7 through 2, had been swept away by a huge tidal wave.

Comparisons of commissarial officeholders in the 1920s as listed in capital city directories with those that appear in service lists (spiski), rank (chin) registers, and professional registers of 1911 through 1916 indicate virtually no holdovers from the high-rank categories in provincial administration. Significantly, in these categories before 1917, in both the central and provincial administrations, one finds concentrated the best-educated, and most experienced of Imperial administrators. Not coincidentally, of course, one also finds the highest concentrations of social elites—landed and landless—who, as we saw in Chapter 2, managed to hold on to their educational advantages and consequent career advantages until the very end of the Old Regime. Representing in themselves so much of what was characteristic of the elites of prerevolutionary society, they were broadly replaced by a junior, much less educated, less experienced population of officials. This was the first large-scale effect on administration of the Revolution of 1917. What it meant in practice was this: not only did the newly emergent specialists, whose fortunes were traced in the preceding chapter, have to make their peace with a lower-class officialdom but, equally, Old Bolshevik intellectuals such as Trotsky and Bukharin were now

---

35. W. M. Pintner, "Evolution of Civil Officialdom," in Pintner and Rowney, *Russian Officialdom*, 223.

The Communist Party and the Soviet Elite

obliged to rub shoulders with what was certainly a green and, quite possibly, even a callow mass.

The term "old," moreover, is used rather more advisedly than is usually the case, and not merely with respect to the Old Bolsheviks. The patterns of age distribution among senior officials and Party members are of some interest and it is frustrating that really satisfactory data on this subject have eluded researchers. In the first place, so far as the administration generally is concerned, it is striking to note that as the 1920s passed, a phenomenon known to later generations of Kremlinologists occurred: the senior officials grew older and older, possibly owing to some reluctance on the part of the political leadership to replace them with the sub-elites, or pre-elites who were waiting, as it were, in outer offices all across the country. According to the sample from Vesiaev et al., 50 percent of senior administrators were thirty-five years of age or older in 1922.[36] In the 1929 population of administrators described by Bineman and Kheinman, 70 percent were above age thirty-five and 28 percent were above age forty-five.[37] These age levels are similar to those of senior civil servants (grade 5 and higher) in the Ministry of Internal Affairs in the 1870s, according to data from Orlovsky.[38]

But what about the Party membership that might be expecting to move into some of these offices? Unfortunately, a direct comparison of the administrative age data with age data for those employed in senior administrative categories from the Party eludes us. One can cite some impressive figures from the Party as a whole, however. According to the 1927 Party census, fully 70 percent of Party members were under age thirty-five; 50 percent were less than thirty. The implication, of course, is not only that Party members who were serving in the state administration were significantly less educated than other senior civil servants but that they were also considerably younger, very possibly eagerly awaiting their next opportunity for a move upward in the administrative hierarchy.

These observations reinforce the conclusion that we are looking at the career and demographic records of two quite distinct, and potentially competitive populations of administrators. Equally, these observations suggest that the lines of competition may have been drawn as early as 1927—perhaps earlier: Bailes dates the distinction between "bourgeois" technical elites and Communist managerial cadres in So-

---

36. Vasiaev et al., *Dannye perepisi*, 51.
37. Bineman and Kheinman, *Kadry*, Table 1.
38. Daniel Orlovsky, "High Officials in the Ministry of Internal Affairs, 1855–81," in Pintner and Rowney, *Russian Officialdom*, 259–60.

viet industry from about 1923.[39] He also remarks that the "organization of personnel and the fulfillment of production goals in industry came to depend on the cooperation of these two groups." Still, competition between them was never far from the surface and as Communists began to take over more and more of the senior industrial administrative positions, the median educational level of the senior administration declined, as one would expect.[40] Later in this analysis I will draw further important implications from the contrast between these two subdivisions of the senior Soviet administration.

The Impact of Proletarianization on the Rank and File

Let us look in greater detail at the experience of social mobility of the party rank and file as revealed by Party statistics. If, indeed, the Party was an important—perhaps, for some, the important—means of upward mobility during the 1920s, we will also be looking at the formation of the first truly Soviet pre-elite.

Official data on Party members, published annually from the middle 1920s, consistently show very high rates of transfer out of the working class and the peasantry (see Figures 4–7). In 1924, 44 percent of Party members were listed as workers by social position but only 18.8 percent as workers by employment. From 1925 on, the data describe relative and absolute representation in the Party of the three main social classes of origin—or pre-Party employment categories—and of the same three classifications considered in current employment. It is evident that in the Party membership from 1925 to 1929 there was a lively rate of transfer out of either the working class or the peasantry, or both, and into the white-collar ranks.

Given that turnover of membership and Party growth rates were high through much of the decade, the evidence suggests that the effects of social transfer were present in a constantly shifting population of Party members. In other words, these people were not simply the same group advancing in Party status and job status from one year to another. Rather, high proportions of Party membership turned over each year, and both the total size and composition of the Party changed considerably during the 1920s.

For example, in 1925 *at least* 325,000 of a total membership of 800,000 Communists were new to the Party *that year*, according to the

---

39. Kendall E. Bailes, *Technology and Society under Lenin and Stalin*, (Princeton, N. J., 1978), 64.
40. Ibid., 64–65.

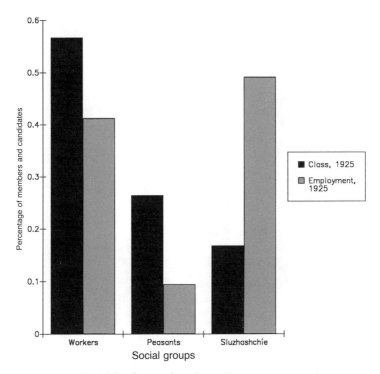

FIGURE 4. Social background and employment status of party members and candidates, 1925. Source: Bol'shaia sovetskaia entsiklopediia, 9:533; T. H. Rigby, Communist Party Membership in the USSR (Princeton, N.J., 1968), 116.

survey of Party history published in the "Great Soviet Encyclopedia."[41] The actual proportion of new members was probably somewhat higher than even that figure suggests, since attrition in the form of resignations or dismissals was typically high among new members. For example, by 1927 at least 30 percent of the 1925 inductees had apparently disappeared from the roles. From 1926 through 1928 the minimum proportion of new members was highly erratic: 26 percent, 6 percent, 12 percent for the three years respectively. Nevertheless, as Figures 4–7 demonstrate, the incidence of transfer out of a social-class category into a new work-class category remained high throughout the period.

By the end of the 1920s the Statistical Department of the All Union Communist Party had gathered volumes of information on the social status of Communists, underscoring the Party's sensitivity to this issue.

41. Bol'shaia sovetskaia entsiklopediia, 9:533; and Rigby, Communist Party Membership, 116.

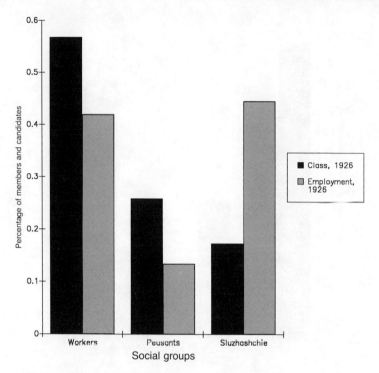

FIGURE 5. Social background and employment status of party members and candidates, 1926. Source: See Figure 4.

But even at the very end of the decade, collected data revealed that the numbers of Communists employed were workers, 727,364; peasants, 189,067; and senior-middle white-collar workers, 472,090; these three employment categories represented about 43 percent, 11 percent, and 28 percent, respectively, of all Communists and candidates in the Soviet Union in 1930. At that time, as Fitzpatrick has shown, a campaign to increase the access of lower-class workers to the training and education necessary for promotion into management careers was well under way.[42] She describes this program of rapid promotion, or *vydvizhenie*, as radically altering the class makeup both of higher educational institutions and of management cadres in the 1930s.

Throughout the 1920s, however, as Rigby has noted, it was Party policy to expand the proportion of members who were workers—workers not by social origin but by employment "at the bench."[43] The distinction was a simple but necessary one to many. The Party's public

42. Fitzpatrick, *Education and Social Mobility*, chaps. 7–9.
43. Rigby, *Communist Party Membership*, chap. 3.

The Communist Party and the Soviet Elite 151

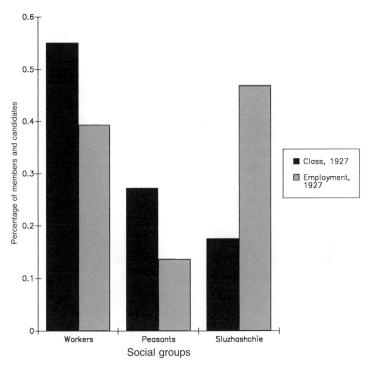

FIGURE 6. Social background and employment status of party members and candidates, 1927. Source: See Figure 4.

image, at least, demanded a high proportion of members who earned their livings as proletarians. In fact, the goal was never achieved, in spite of a major purge that attempted to weed out socially and ideologically undesirable members and in spite of major campaigns throughout the decade to expand membership and to recruit workers.

The fact of the failure of this Party goal should be carefully noted. It was not merely an indication of the hierarchy's misunderstanding of why people would join the Party; it actually seems to have been a measure of the inability of the leadership to control who was entering, what happened to them after entering, and how those who entered could use their newly won influence in their own behalf. J. Arch Getty provides many examples of breakdowns in linkage in the 1930s between the Party center and its announced policies on the one hand, and behavior in the periphery on the other.[44] Trotsky, referring to the 1920s,

44. J. Arch Getty, *Origins of the Great Purges: The Soviet Communist Party Reconsidered, 1933–1938* (New York, 1985), chap. 3, and "Party and Purge in Smolensk: 1933–1937," *Slavic Review* 42, no. 1 (1983), 60–79.

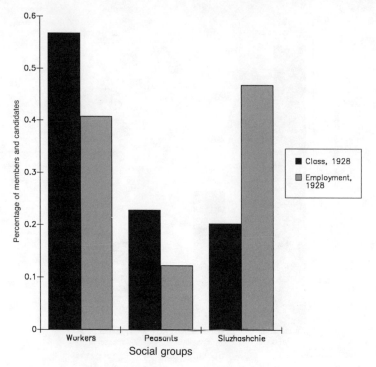

FIGURE 7. Social background and employment status of party members and candidates, 1928. Source: See Figure 4.

said that with Stalin's help the bureaucracy conquered the Bolshevik Party "with its own social weight."[45] No doubt he was being ironic; nevertheless, his assertion seems completely plausible. But it was evidently also true that the Party was acting willy-nilly as a magnet to careerists, to the ambitious and the upwardly mobile, who could be as easily found among workers and peasants as anywhere else, not only in the 1930s but in the 1920s too.

We have already noted the high turnover in Party membership, especially during the early years. It may well be that the motivation of workers to continue with time-consuming Party activities was not as strong as it was for other social groups. During this era, moreover, the membership tended to be young, with the Party primarily attracting individuals in their twentys and thirtys. Rigby confirms this characteristic for the Party generally, and D.J. Male suggests that it was so for

45. Trotsky, *The Revolution Betrayed*, 94.

TABLE 28.
Social origins of Communists in state and white-collar positions, 1927

Background	N	Civil agencies	White-collar roles	Proportion in Party
Workers	599,111	42%	44%	56%
Peasants	195,608	12	13	18
White-collar employees	237,912	42	40	22
Others	29,229	3	3	3

Sources: For workers, white collars, and others, E. Smitten, *Sostav Vsesoiuznoi Kommunisticheskoi Partii (bol'shevikov)* (Moscow, 1927), 23, Table 8; 29, Table 11; 51, Table 25; 52, Table 26; for peasants, Central Committee, All-Union Communist Party (bolsheviks), Statistical Division, *Sotsial'nyi i natsional'nyi sostav VKP(b)* (Moscow, 1928), 100, Table 16.

the peasantry in particular.[46] Such groups would have been likely to contain a higher proportion of workers than of older, more established, more seasoned Party members who had had ample time to achieve career advancement. Thus it seems likely that high attrition was concentrated among new members, young members, and workers—simultaneously—and that turnover may thus have been an important reason for the failure of the Party policy of proletarianization.

The most important reason for the failure of proletarianization, however, is the one most relevant here. Throughout the 1920s the Party appears to have recruited high proportions of individuals who were workers and peasants *by social origin*, but the positions these individuals held or had gained by the time of the Party census were by no means necessarily parallel to their social origins. Of the 1,061,860 members and candidates registered (see Table 28) in the Party census of 1927, well over half were of working-class origin, but only 32 percent were workers by profession. Similarly, 18 percent of members and candidates were peasants by origin, but only 15 percent were peasants by occupation.[47] Clearly, though the intentions of recruiters may have been excellently in line with Party policy, their recruits did not always live up to expectations. They chose to leave the bench. As Table 28 shows, of all the Communists in civil agencies, 42 percent were workers by origin; of those in all white-collar roles, 44 percent.

In order to understand what was happening to Party members who were no longer in working-class (including peasant) employment, one should look rather more closely at both the overall Party membership patterns in the 1920s and, in particular, at the patterns of variation in

46. Rigby, *Communist Party Membership*, 353; D. J. Male, *Russian Peasant Organisation before Collectivisation: A Study of Commune and Gathering, 1925–1930* (Cambridge, 1971), 69.
47. Central Committee, All-Union Communist Party, *Sotsial'nyi i natsional'nyi sostav*, 52, Table 2; 100, Table 16.

the non-peasant and non-working-class employment categories of Party members. After all, while there is substantial evidence of status change among Party members, it is by no means clear what the nature of that change was. How many of the workers who departed from working-class status became peasants? And were the peasants moving up? If so, was it one notch to working-class status or two notches to white-collar status?

These questions are partially answered by Table 29, which summarizes employment patterns of workers, white-collar employees, and peasants in rural Party cells. What happened to the 92,000 peasant-class rural Communists who were not working as peasants? The answer is that the vast majority of them, about 56,000, found white-collar positions, indicating a massive jump upward for some, at least—not merely to indoor work but to sit-down, "gentleman's" work.

A similar pattern holds for workers, whose overall employment status is also given in Table 29, together with that of the other major class categories in 1927. For working class members *overall*, as for peasant members overall, the prospects of upward mobility were good. At some point at least a third of each group had found it possible not only to move out of their social class of origin but to move up into the white-collar employment bracket or into the student category, which should ultimately have conveyed white-collar status. Party membership was apparently associated also with the retention of their white collars by persons from the white-collar *class*: almost four-fifths managed to retain their relatively elevated status by gaining some form of white-collar employment.

One of the reasons for the enormous increase of the working-class presence in schools and in white-collar offices after 1928 was that the number of places in these categories increased significantly. The same observation may be made for the 1920s, although as Sternheimer has shown, the problem of estimating the actual size of the early Soviet state bureaucracy is complex and uncertain.[48] The expansion of offices was not as enormous as it would be in the 1930s, as a comparison with Fitzpatrick's findings indicates, but it was large all the same.[49] And the kinds of changes—the offices and statuses involved—are significant.

According to Smitten, government service personnel and personnel in the cooperative programs in 1924 totaled about 1,866,000.[50] By 1926 the figure had increased by 25 percent, to 2,330,000. In the same period the number of Communists in such positions had increased from

---

48. Sternheimer, "Administration for Development," 320ff.
49. Fitzpatrick, *Education and Social Mobility*, 238ff.
50. Smitten, *Sostav*, 53.

TABLE 29.
Employment status of principal class categories of Communists and candidates, 1927

Employment		Social class[a]			
		Worker	Peasant	White collar	Other
Workers at the bench	%	53.3	5.9	3.9	9.5
	N	(319,326)	(11,541)	(9,279)	(2,777)
Peasants working the land	%	1.7	52.8	0.9	2.2
	N	(10,185)	(103,281)	(2,141)	(643)
White-collar and social workers	%	30.8	28.7	78.9	37.1
	N	(184,526)	(56,139+)	(187,713)	(10,844)
Junior service personnel	%	2.7	1.7	3.4	3.7
	N	(16,176)	(3,325)	(8,089)	(1,082+)
Students	%	4.9	6.3	5.4	11.5
	N	(29,357+)	(12,323)	(12,847)	(3,361)
Unemployed persons	%	4.6	1.8	4.7	8.4
	N	(27,559)	(3,521)	(11,182)	(2,455)
Persons unable to work	%	0.5	0.1	0.4	0.9
	N	(2,966+)	(196)	(952)	(263)
Domestic workers	%	0.3	0.3	0.6	18.2
	N	(1,797)	(587)	(1,427+)	(5,320)
Others	%	1.2	2.4	1.8	8.5
	N	(7,189)	(4,695)	(4,282)	(2,484+)
TOTAL	%	56.4	18.4	22.4	2.8
	N	599,111[b]	195,608	237,912	29,229

Source: E. Smitten, *Sostav vsesoiuznoi kommunisticheskoi partii (bol'shevikov)* (Moscow, 1927), 23, Table 8; 27. Table 8 excludes members in the armed forces and those living abroad.
[a]The source provides only row totals, column totals, and percentages to one decimal place; all figures in parentheses are calculated from the data.
[b]The figure given in the source is in error; the correct figure can be found in Central Committee, All-Union Communist Party (bolsheviks), Statistical Division, *Sotsial'nyi i natsional'nyi sostav VKP(b)* (Moscow, 1928), 34, Table 9.

256,300 to 395,000, over 50 percent. Moreover, it is evident that there were certain administrative positions—such as those with local or regional soviets—upon which peasants and workers appear to have had first claim, perhaps for political reasons, perhaps for reasons of location.

Striking as these numbers are, they do not satisfactorily interpret the data already presented. The argument to be confronted now is that membership in the Party was a vital means of upward social mobility for a certain segment of Soviet society who were both lower-class and relatively uneducated. An important corollary to the argument is that mobility occurred at all levels, beginning in 1925 or earlier. The principal indication of the elevation of status has been the attainment of white-collar employment. In order to clarify the impact of Party membership on mobility, it is helpful to consider two related questions. To what extent did the civil administration offer opportunities for social advancement that were independent of those opened up by the Party? To what degree were the same opportunities open to urban and rural (non-worker) Party members?

Tables 30 and 31 provide comparisons between the types of white-collar positions held by members of the Communist Party and by civil servants generally. The comparisons are subdivided by social class and type of white-collar position. The reader should keep in mind that the Party data are from early 1927 and the civil administration data from 1929—a difference of only two years but a potentially important two years during which there was a major shakeup of Party membership and of administrative officeholding.[51] In spite of this difficulty, the contrasts are interesting and important.

The data underscore an important difference in mobility opportunities among the lower classes: Party membership was advantageous to upwardly mobile workers, but work in civil administration was advantageous to peasants. Among Party members with white-collar jobs, individuals from working-class backgrounds consistently outnumbered peasants, as they did in the Party generally; it should be noted, however, that the workers' advantage over peasants in white-collar employment was somewhat greater than their proportionate advantage over peasants in Party membership generally. Workers were advantaged in selection into the Party, but they were even more advantaged in terms of access to white-collar employment. According to Smitten, whose data were summarized in Table 29, workers by social origin constituted about 56 percent of all Party members, and peasants accounted for 18 percent

---

51. See, e.g., J. V. Stalin, "On Achievement of the Next Tasks of the Struggle Against Bureaucratism," *Pravda*, 12 June 1929.

TABLE 30.
Background of Party members in white-collar positions, 1927

Position	N^a	Calculated totals	Social background[b]			
			Worker (43.8%)	Peasant (13.7%)	White collar (39.7%)	Other (2.9%)
Sluzhashchie	525,604	(525,780)	(230,058)	(71,787)	(208,649)	(15,286)
High and medium	438,832	(439,222)	(184,526)	(56,139)	(187,713)	(10,844)
Low	28,777	(28,671)	(16,176)	(3,325)	(8,089)	(1,081)
Students	57,995	(57,887)	(29,356)	(12,323)	(12,847)	(3,361)

Source: E. Smitten, *Sostav vsesoiuznoi kommunisticheskoi partii (bol'shevikov)* (Moscow, 1927).
[a]From Smitten, 24.
[b]Calculated from Smitten, 23, Table 8. The calculated totals are inaccurate as a result of rounding.

TABLE 31.
Commissarial staff distribution by office level and class, 1929

Office level	Class of origin							
	Worker (N, %)		Peasant (N, %)		White collar (N, %)		Other (N, %)		Total (N)
All Union and Republic	3,985	17.3	1,969	8.5	15,999	69.4	1,101	4.8	23,054
Oblast, okrug, raion	19,303	19.1	34,606	34.2	43,594	43.1	3,647	3.6	101,150
Total	23,288	18.7	36,575	29.4	59,593	48.0	4,748	3.8	124,204

Source: Ia. Bineman and S. Kheinman, *Kadry gosudarstvennogo i kooperativnogo apparata SSSR* (Moscow, 1930), 86. Table 1.

The Communist Party and the Soviet Elite 159

at the time of the 1927 Party census.[52] As Table 30 indicates, however, peasants made up only 13.7 percent of Party members who held white-collar jobs, while workers constituted almost 44 percent of Party white-collar employees. On the other hand, we see from Table 31 that peasants accounted for about 30 percent of civil officials at all levels of the civil administration without regard to Party membership; the comparable figure for workers was only 19 percent. The tendency for all former workers and peasants (not just Communists) to be concentrated at the regional and subregional administrative levels is illustrated in Table 31. This table also underscores the tendency for workers to have had greater access than peasants to those 6,000 or so positions that were held by the lower classes in the central (All Union and Union Republic) government.

Proletarianization without Education

Lest the import of these figures be obscured, it is useful to keep in mind that they describe a political organization, the Party, extraordinarily different from anything that had existed in Russia a mere decade earlier. At the same time, in spite of the obvious motivation for distorting or even fabricating data on social origins, it remains true that there existed by the second half of the 1920s a socioeconomic group, the white-collar workers (sluzhashchie), undreamed of even a scant seven or eight years earlier. It was *the* functional administrative category, and not only was it controlled by the Party, but its membership included a large proportion of men and women of working-class and even of peasant origin. Taken as a measure of the degree to which the Revolution of 1917 could be called social as well as political, the data on Party membership and white-collar origins seem to proclaim that enormous advances had been made in ten years' time by thousands of Russia's poor and uneducated. The data bespeak a social revolution of great dimensions, just as they argue a political revolution of similar size.

These results were not, of course, in direct accord with Party policy; the goal of 50 percent active worker membership continually receded because of the upward mobility that Party membership conferred. Although policy was to make "every second Communist" a bench worker, events and other policy worked at cross-purposes with this objective. Still, it is not possible to say that the results were *contrary* to policy. In fact, the Party was somewhat ambiguous on the point of what should happen to workers once they entered the Party. Earlier in the 1920s it

52. Smitten, *Sostav,* 23.

had announced itself in favor of transfers of status: "The chief task standing before the Party in connection with the incorporation of over 200,000 new members consists of arranging to draw them into state work.... This should not be hampered by the lack of training of the new members, and in particular by their non-completion of a propaganda course."[53]

Thus, it is hard for us to say which policy was supposed to prevail. What is important here, however, is that before 1930—even before 1928—workers and peasants *did* advance in Soviet society in substantial numbers, confirming for themselves the promises of the revolution. The fact that they did so largely without education seems to confirm the political aspect of their power. The fact that they did so at all makes the Soviet revolution unique in European history, unless one is prepared to argue that the creation of the Party and the structuring of a new servitor class were events without either political or social significance. Nowhere before had the reins of power changed hands so swiftly and so decisively. This becomes especially evident if we keep in mind the social-demographic shift in the makeup not only of the administrative apparatus broadly conceived but of the senior and political levels, the Central Committee and the Council of People's Commissars. The fact that peasants and workers advanced also seems likely to have placed the Party elite (coming more and more from the lower classes itself) under pressure to recognize the potential influence of workers and peasants and their demands for further advancement (I shall return to this point.

Table 32 summarizes lower-class access overall to institutions that either conferred or promised upward social mobility before the end of the 1920s. To be sure, there is ample evidence of the holding power of individuals who were members of the traditional white-collar classes, as we saw in Chapter 4. But the striking fact is that even before the crash programs for education and economic advancement at the end of the 1920s could have had significant impact, peasants and workers were heavily represented in a broad range of white-collar institutions—this in a country where, ten years before, such institutions were not always hospitable to the small middle and professional classes, let alone the lower classes. Moreover, the institutions that served as means for this lower-class upward mobility were the politically most powerful, the Party and the state bureaucracies.

Data from the 1927 Party census also make it possible to say what roles, in general, were being played by Communists holding white-

---

53. Institute of Marxism-Leninism, *KPSS: v rezoliutsiiakh i resheniiakh s"ezdov, konferentsii i plenumov TsK* (Moscow, 1954, 1:824.

TABLE 32.
Communist and non-Communist access to white-collar status, 1927–29

Institution	Class of Origin						Total
	Worker		Peasant		White Collar		
	N	%	N	%	N	%	
Tertiary schools	(40,589)	25.4	(37,713)	23.6[a]	(81,498)	51	159,800
Civil administration	152,095	18.4	198,189	24.0	415,042	50.0	825,100
Communist Party Members with white-collar status	(184,526)	43.1	(56,139)	13.1	(187,713)	43.8	(428,378)
	(200,702)[b]	44.0	(59,464)[b]	13.0	(195,802)[b]	42.9	(455,968)[b]
Members with white-collar and student status	(213,882)[c]	44.3	(68,462)[c]	14.2	(200,560)[c]	41.5	(482,904)[c]

Sources: Sheila Fitzpatrick, *Education and Social Mobility in the Soviet Union, 1921–1924* (Cambridge, 1979), 189 (Table 2), 197, 106–7 (tertiary schools); Ia. Bineman and S. Kheinman, *Kadry gosudarstvennogo i kooperativnogo apparata SSSR* (Moscow, 1930), 86, Table 1 (civil administration); E. Smitten, *Sostav vsesoiuznoi kommunisticheskoi partii (bol'shevikov)* (Moscow, 1927), 23 (Table 8), 24 (Communist Party). Data in parentheses are calculated.
[a]Includes "others": i.e., bourgeoisie and additional non-worker, non-white-collar groups, usually less than 5%.
[b]Includes "junior service" category: cf. Smitten, Table 8.
[c]Includes only executives, middle management, and students; see cols. 6 and 8 in Smitten, Table 8.

TABLE 33.
Communists in administrative positions, 1927

Administrative job type	N	%
Soviet administrative	49,878	13.5
Industry	49,313	13.4
Cultural-political education	41,434	11.2
Trade	34,381	9.3
Professional	28,213	7.6
Communist Party	27,607	7.5
Transport	24,008	6.5
Courts and prisons	22,679	6.1
Cooperatives	21,712	5.9
Health and welfare	17,652	4.8
Credit and finance	12,513	3.4
Land	8,297	2.2
Communist Youth	8,255	2.2
Communications	5,574	1.5
Press	3,067	.8
Volunteer societies	3,063	.8
Construction	2,105	.6
Planning-control	802	.2
Other	8,662	2.3
Total	369,215	99.8

Source: Central Committee, All-Union Communist Party (bolsheviks), Statistical Division, *Kommunisty v sostave apparata gosuchrezhdenii i obshchestvennykh organizatsii* (Moscow, 1927), 55, Table 1.

Note: Data do not include the Ukrainian SSR; Kamchatka, Vilnius, and Kolyma okrugs; or Pri-Amur raion. Job types are ranked in order of frequency.

collar positions. These data (Table 33) are unfortunately incomplete, but they do provide some indications. Nearly half (47.4 percent) of all Communists counted in these data who were employed in white-collar roles were engaged in one of four areas: soviet (that is, mainly local or regional government) administration, cultural-political education, industrial administration, or trade administration. The importance of state employment is underscored by the roles identified. The vast majority of these roles were in one or another of the institutions that were part of the state apparatus: the soviets, educational institutions, courts and judicial administration, professional organizations, or substantial proportions of industrial, trade, and transport organizations. These data fit with Smitten's observations that of a total Party membership of about 1,000,000 in 1927, some 439,000 were employed by the state directly in the state apparatus or indirectly in such "social" or "economic" institutions as credit organizations and cooperatives.[54]

Interestingly, as shown in Table 28, the 1927 Party census provides

54. Smitten, *Sostav*, 24.

The Communist Party and the Soviet Elite   163

a breakdown of state employment patterns of Communists by class as contrasted with their white-collar roles generally, making it possible to estimate the proportions of Communists in each class who were employed in any state-related positions. Again, the data underscore the importance of the state as a white-collar employer; they also underscore the gains made by workers and peasants in using the state and related administrative apparatuses as ports of entry to elevated work status.

The role of personnel who were *socially classified* as white-collar is also emphasized in Table 28. This group is the only social class whose proportion in the state apparatus was actually greater than its proportion among Party members holding white-collar positions generally. One cannot escape the impression that the elevated share of roles in the state apparatus had been sustained at the expense of workers and peasants who were Party members but who had not as yet found the means of entering a higher class status. Even among Party members, competition for desirable employment seems to have concentrated along lines of class, augmented by education and experience. So long as the socially elevated classes held onto their jobs, and so long as there was no "natural expansion" of jobs (in Fitzpatrick's phrase), they stood in the way of additional worker and peasant entry into the state system or up through the state system.[55]

It is in this context, I believe, that the full meaning of the age data cited earlier in this chapter can be interpreted. During the 1920s the tenacity of those in high-status administrative jobs—a tenacity that was reflected in their advancing age as a group—must have slowed promotion from lower levels. Given the class patterns we have already seen, this same phenomenon must also have meant a slowing in the pace of advancement by peasants and especially workers to higher administrative status. No doubt the survival of the white-collar social class among Party members should be explained in the same way as the "holdover" phenomenon (described in Chapter 4) and the white-collar survival in the civil service at large. These people possessed experience and education that were needed by both Party and state. One may ask, however, whether such an explanation was satisfactory to thousands of workers and peasants who saw that want of education was no obstacle to their peers in many other state and Party positions.

The results of this analysis correspond well with observations on structural change in the administrative apparatus during the postrevolutionary period. The connection between noble landholding and noble administrative dominance, as noted earlier, seemed to ensure rapid and dramatic changes in regional and local administration during

---

55. Fitzpatrick, *Education and Social Mobility*, 17.

the revolution. One has only to recall the principal characteristics of provincial civil service, provincial landholding, and provincial political institutions just before 1917 in order to grasp the immensity of the shift that actually occurred. By the mid-1920s, workers and peasants were heavily represented in the civil administration, especially at the local and regional levels and in certain job categories. Where they had direct physical access to administrative positions—in the districts and villages—and where they had comparatively limited competition from the white-collar social class, workers and peasants made substantial inroads, holding tens of thousands of administrative offices.

That the holders of these offices were not well educated, we have already seen. Educational qualifications declined dramatically as distance from the center increased and as the proportion of lower-class officeholders increased. Thus, the change in status from worker or peasant to administrator *preceded* any change in qualifications for the workers or peasants involved. This view is contrary to the findings of L. M. Chizhova, who suggests that training and lower-class advancement were closely related even in the 1920s.[56] To me it seems, on the contrary, that mobility independent of education is exactly what one would expect from members of social classes who typically attach great importance to job status and who think of education merely as a means to gaining higher status. If education was not necessary—and in general, it evidently was not—then why bother? Thus, if there was a revolution from above after 1928, it seems likely that it consisted at least in part in the imposition of educational standards on people who theretofore had escaped them but who had nevertheless enjoyed some career or job advancement.

Yet as already noted in Chapter 4, it is the case that white-collar employees from the former white-collar classes were not randomly distributed throughout the state and related apparatuses. But one can now discern that their distribution among officeholders was a function of something more than the factors identified earlier—that is, avoidance of politically sensitive and of functionally novel positions. As the data on the broad reaches of the civil administration and Party white-collar employment show, the *sluzhashchie*—accounting, no doubt, for a high proportion of the tsarist holdovers—were concentrated both at the national levels of administration and in the specialist roles where their education was indispensable. As Sternheimer notes, they certainly did not dominate the administrative apparatus (because of changing struc-

---

56. L. M. Chizhova, "Vydvizhenie, 1921–1937 gg.," *Voprosy istorii KPSS*, no. 9 (1973), 114–26, and "K istorii pervykh mobilizatsii partiinykh rabotnikov v derevniu," *Sovetskie arkhivy* 2 (1968), 69–74.

ture and changing personnel, it was becoming distinctly "soviet" in its overall characteristics), but at the same time, even several years after the revolution, they were evidently vital in some offices. Thus, part of the explanation for the post-1928 policies of worker advancement must indeed be that workers and peasants needed to be trained to assume the roles that had been dominated by the former lower echelons of the old serving classes, as well as to fill the thousands of new positions of the same type that industrialization and collectivization were bound to open up. Certain facts, however—that it was the lower classes who were poised to take advantage of these policies, that the policies were tilted in their behalf, and that the debates of the 1920s over these policies often fixed on their potential impact on the lower classes—must all be explained, it seems to me, not merely in terms of Communist ideology or of Stalin's personal choice but also in terms of the enormous influence that these lower classes already possessed in both the Party and the civil administration.

Whatever may be the finer points embedded in the explanation of the policies of the later 1920s, one should not lose sight of the conclusion that sociologically and organizationally, the system as a whole—both Party and state—was the product of major changes wrought in a very short time.

The Impact of Proletarianization: Some Caveats

The changes that the data suggest in lower-class status and in the incumbencies in political and administrative roles were often undoubtedly more apparent than real. Lorimer, for example, shows that many positions held by peasants in the civil administration were considered secondary positions by individuals whose principal identification was with farming. Lewin, Male, Shanin, Carr, and Davies, moreover, all find that rural Party members and administrators frequently occupied ambiguous positions and were far from respected or accepted by the rural communes.[57] Indeed, the overall importance of the administration at the regional and subregional levels has been placed in considerable doubt by both Male and Davies, who point out that very often the real successor to tsarist local administration was not the local civil admin-

---

57. For data on secondary employment of peasants in public service, see Frank Lorimer, *The Population of the Soviet Union: History and Prospects* (Geneva, 1946), 74, Table 28. For additional discussion of peasant administrative roles see Moshe Lewin, *Russian Peasants and Soviet Power: A Study of Collectivization*, (New York, 1968), 119–26; Male, *Russian Peasant Organisation*, 127ff.; Teodor Shanin, *The Awkward Class* (Oxford, 1972), 163–66; E. H. Carr, and R. W. Davies, *Foundations of a Planned Economy* (New York, 1971), 2:179ff.; R. W. Davies, *The Socialist Offensive: The Collectivisation of Soviet Agriculture, 1929–1930* (Cambridge, Mass., 1980), 1:53.

istrators but the peasant communal gathering (skhod). Party data—from Smitten, for example—also indicate that a considerable proportion of Party members maintained "ties" with the land, presumably with the commune.[58]

The extent to which white-collar positions really produced a change of status in the cases cited here, and the quality of the change, may well be debated. Undeniably there were substantial variations in the real effect of the status change implied by a move into a white-collar position. Many of those who enjoyed white-collar status in rural areas, moreover, were "outsiders," not peasants, as Shanin shows.[59] Nevertheless, tens of thousands of workers and peasants did hold low-level administrative positions, derived income and status from some of them, and were thereby identified by family and peers as participants in the state and Party systems.

Variation in Access to Party Positions

Other factors also modified job status or social classification in this large country. Considered as an institution whose organization stretched across parts of two continents, the Party was understandably far from homogeneous. For example, Party census data as reported by Smitten indicated considerable variation in the level of skills or qualifications that members and candidates brought to their jobs. Not unexpectedly, members employed as agricultural labor were generally classified as possessing few ("semi-skilled") or no ("unskilled") job qualifications.[60] Only some 11 percent of this group were skilled workers, as contrasted with 60 percent of industrial workers.

It is possible to look at overall geographical distribution of workers arranged by their qualifications and to match these patterns with the social and economic characteristics of the major geographic regions identified by Lorimer from the 1926 general census.[61] Such comparisons may be reduced to the simple statistics found in Table 34, in which the populations of both skilled and unskilled workers are correlated with specific regional characteristics. The levels of correlation are not uniformly high, but what is striking is the consistency of the direction of the correlations.

Among other things, the comparison in Table 34 suggests that unskilled workers who were members or candidate members of the Communist Party tended to be concentrated in the same geographic areas

58. Smitten, Sostav, 41.
59. Shanin, Awkward Class, 183–85.
60. Smitten, Sostav, 40.
61. Lorimer, Population, Table 26.

TABLE 34.
Skill level of worker-Communists correlated with socioeconomic characteristics of their area of residence, 1926–27

Workers	Population					(7) Proportion unemployed
	(3) Per square kilometer	(4) Urban	(5) Literate (age 10+)	(6) Income		
(1) Skilled	.43	.46	.31	.72		.46
(2) Unskilled	−.45	−.15	−.59	−.52		−.49

Note: Product-moment (Pearson) correlation of seven variables selected from a data set of fifty-six Soviet economic, geographic, and demographic characteristics. The seven derive from the following sources: (1) and (2) from Central Committee, All-Union Communist Party (bolsheviks), Statistical Division, Sotsial'nyi i natsional'nyi sostav VKP(b) (Moscow, 1928), 133, Table 13; (3), (4), and (5) from Frank Lorimer, The Population of the Soviet Union (Geneva, 1946); 67, Table 26 (data for 1926); (6) from Lorimer, 76, Table 29 (data for 1926); (7) from Lorimer, 71, Table 27.

where we would expect to find the rural members—in the heavily agricultural, often poor, and overpopulated provinces of the Russian Republic and the Ukraine, in Dagestan, Kazakhstan, Kirghizia, and Uzbekistan. Even though there were high proportions of "workers" in the Party or in the state administration in such areas, the chances are good that they were little better prepared for their roles than were the peasants. In spite of the high proportion of white-collar roles they filled, the commitment of these individuals to the "working-class" values of the Party and to its goals was probably low.

Similarly, according to the 1927 Party census, these same relatively agricultural regions tended to be high in Party personnel who were merely candidate members.[62] Many of the regions where there were high proportions of unskilled workers also had exceptionally high ratios of candidates to Party members, reflecting not only the youthfulness of the Party but probably also its instability and high turnover.

The Party data further show that the tendency to leave the working class for non-working-class employment was considerably stronger in rural areas than elsewhere. Table 35 presents this phenomenon in general for rural areas throughout the USSR. Of the 49,510 working-class Communists in rural areas who were performing in roles other than that of worker, a huge 30,562, or 62 percent, were in white-collar jobs. This figure should be compared with the 30.8 percent of working-class Party members filling white-collar positions generally. For the less skilled workers in rural areas, the proportion in white-collar roles was 46.5 percent.

62. Central Committee, All-Union Communist Party, Sotsial'nyi i natsional'nyi sostav, 23.

TABLE 35.
Mobility patterns of Communists in rural areas, 1927

Social class	All Communists	Workers and agriculture laborers	Peasants	White-collar employees	Junior and service personnel	Handcrafters and artisans	Unemployed	Others
Worker	65,691 (25.6%)	16,181 (68.5%)	9,528 (8.6%)	30,562 (29.8%)	2,116 (40.1%)	906 (35.8%)	5,214 (58.9%)	1,184 (38.9%)
Peasant	149,734 (58.3%)	6,429 (27.2%)	99,540 (89.6%)	36,989 (36.1%)	2,147 (40.6%)	1,544 (61.1%)	1,845 (20.9%)	1,240 (40.7%)
White collar	41,346 (16.1%)	1,022 (4.3%)	1,982 (1.8%)	34,837 (34.0%)	1,018 (19.3%)	78 (3.1%)	1,789 (20.2%)	620 (20.4%)
Total	256,772	23,633	111,051	102,389	5,282	2,529	8,849	3,045

Source: Central Committee, All-Union Communist Party (bolsheviks), Statistical Division, Sotsial'nyi i natsional'nyi sostav VKP(b) (Moscow, 1928), 93, 100, 103, Tables 11, 16, 19.

Table 35 also indicates the difference that Party membership made in terms of upward mobility for workers as contrasted to peasants. Although peasants who held white-collar positions in the countryside far outnumbered members of any other class, when one considers only Party members, the advantages accorded to even relatively unskilled workers through Party membership and Party influence become quite evident.

It is in the context of the relative importance of Party membership in rural places, perhaps, that one can most usefully speak of a struggle for control of the countryside before the onset of the collectivization of agriculture. With the collapse of the tsarist system of rural administration, peasants apparently moved to fill the local power vacuum and perhaps to ensure favorable control over their own economic and political fate. Party policies of the mid-1920s aiming to advance the cause of worker status appear to have challenged peasant dominance of the countryside by favoring workers, when and where there were workers to favor. But the data collected by Bineman and Kheinman, in the decade's last major study clearly suggest that the policy had failed if its goal was to achieve worker (even unskilled worker) dominance of the countryside. Instead, the countryside was administratively dominated by the peasantry, and these *peasants* were becoming Communist in considerable numbers.[63]

Western students of Soviet political development have sometimes puzzled over the tendency for Soviet *apparatchiki* of the later 1930s to have been peasants by background instead of workers. In fact, the reasons are not hard to discover. If one examines the social composition of the lower administrative cadres—the pre-elite or sub-elite—at the end of the 1920s, it is clear that the odds were overwhelmingly in favor of the future development of an administrative management predominantly lower-class in origin. The proportion of peasants in the pre-elite, moreover, was sufficiently high to guarantee that even if only a small percentage of that class achieved significant promotion, their number would be considerable.

Collectivization presumably opened up many new administrative positions, in both the Party and the civil administration, to peasants willing to cooperate with the program—although as Davies demonstrates, collectivization relied heavily on thousands of urban Party members who went to the countryside and stayed there.[64] But even earlier (according to the Party census of 1927) there were almost as many peasant secretaries of Party cells as worker secretaries (12,500 to

---

63. Bineman and Kheinman, Kadry, chap. 5.
64. Davies, Socialist Offensive, chap. 5.

12,800). These figures confirm the "structural" advantages for those peasants who wished to improve their social status by taking administrative positions.

Another way of thinking about the influence of peasant administrators even before collectivization is to refer once more to Table 35, which gives the numbers of each social class that composed the white-collar workers in rural Party cells. Of more than 102,000 white-collar members of rural cells, about 30 percent were workers by background and 34 percent were originally from the white-collar class. But the highest proportion, 36 percent, consisted of administrators who had started out as peasants. Even if, as Trotsky asserted, many of these peasants were *kulaks*, or well-to-do farmers, who were using their positions to safeguard their own economic interests and those of others like them, the evidence is nevertheless clear that peasants were a major component of the postrevolutionary pre-elite and sub-elite, whatever the Party may have intended.

Lower-Class Mobility and Administration

In the late winter of 1927, the French intellectual Pierre Pascal observed in his diary: "A decision published today (in [the trade union paper] *Trud*) says that only "the children of workers and white-collar workers" may be registered in the employment roles of the Employment Office. But what if the son of a merchant, not to mention a peasant, should wish to work in a white-collar position? A kind of law of heredity is being created here."[65] Diarists often see some of the truth and, nearly as often, report some of it accurately. Pascal was certainly not alone in predicting that the policies of the 1920s and early 1930s would produce a new bureaucratic caste in Russia, but the data examined here do not appear to confirm either that the "regulation" of 1927 was effective or that Pascal's fears were borne out. Merchants' children may well have been excluded, as were the unreconstructed children of other prerevolutionary upper classes. Peasants as well as workers, nevertheless, managed to move into white-collar roles in such large numbers that the notion of the emergence of a new hereditary elite seems difficult to sustain.

I propose that the educational policies of the late 1920s and early 1930s were measures taken by the political leadership in self-defense rather than bold measures carefully crafted to secure a social revolution. The social revolution was already well under way. Its future was assured before the end of the 1920s: a large proportion of the total civil

---

65. Pascal, *Russie 1927*, 88.

administrative workforce consisted of workers and peasants, and those tens of thousands were in a good position to defend their new-won status and to demand further advancement.

The government, if anything, was in danger of drowning in a sea of bureaucrats who were the heirs of a vast social revolution. With little education and experience, these thousands of administrators were unlikely to contribute to rapid social and economic development, however their political masters should decide to achieve it. Under such conditions, midcareer crash education programs were not only desirable but imperative—if not from the viewpoint of the sub-elite and the would-be pre-elite, then certainly from the viewpoint of those politicians who were committed to Soviet development.

In all, by the end of 1920s, administrators of all kinds accounted for about 10 percent of the total workforce, according to Lorimer.[66] The civil administration examined here made up more than 70 percent of Lorimer's total administrative cohort. The workers and peasants within that group—about 200,000 in all subcategories—constituted a potent force, demanding confirmation of revolutionary gains and opportunities for further gain. Whatever the policies of the Party toward social mobility or education, given the strong position of lower classes in the administrative workforce, the odds were simply in favor of their advancement. Such advancement to higher, even the highest, posts would have to include not only workers but a rather large number of peasants and unskilled, predominantly rural workers, as well.

Russia's peasants and unskilled workers were the lumpenproletariat and petty bourgeois bogeymen of Marx's and Lenin's and Trotsky's nightmares. Together with the skilled workers favored by Marxist ideology, however, they thrust themselves forward as the new raw material of the Soviet elites and sub-elites. Ultimately, because of their vast numbers, they became much more important than the skilled workers.

Proletarianization and Political Control

It now remains to evaluate these conclusions as they bear upon the question of political control—specifically, to probe the implications of Soviet administrative development during the 1920s for the concept that government administration, however large, powerful, and expert, is the servant of the political elite. One way of highlighting this problem is to ask a prior question: If the Soviet political elite of the 1920s had tried to predict important characteristics of the Soviet bureaucracy as it developed in the twenties, how successful would it have been? The

66. Lorimer, *Population*, App. Table A, 1928.

relevance of the question is obvious. If the leadership could not predict the impact of its policy directives on the administration, then control in the sense of conscious direction would be impossible.

One may ask, for example, whether Lenin's "State and Revolution" and "Will the Bolsheviks Retain State Power?" offered good predictions of the future development of the Soviet administrative apparatus. Even the most serious and extended Western analyses of these works depict Lenin as being far wide of the mark so far as questions of structure and the development of operations were concerned.[67] Lenin's personal frustration with attempts to predict the specific outcome of specific political directives to units of the administration—the result of only a few years' work with the state administration—is amply documented, and his favored imagery of machines failing to respond to their drivers' control has already been recalled.

A major goal of Party control after Lenin's death was frustrated in what seems to have been a totally unpredictable way. The goal of a Party membership consisting of 50 percent or more "workers at the bench" eluded the leadership throughout the 1920s. The failure was not for want of trying, however; large numbers of workers were inducted into the Party. Nor did the political elite harbor some deep, hypocritical desire to frustrate worker advancement. The problem was simply that, workers who became Party members ceased being workers at the bench—so consistently, as I have argued above, that upward mobility seems to have been the workers' real objective. The failure of Party policy was, in fact, the measure of workers' success in achieving advancement.

If Lenin's frustration with problems of prediction and control was the result of conflict or contradiction between organizational behavior and political goals or objectives, the failure of the worker recruitment program was more fundamental. It seems to have been the product of misguided interpretations of the impact of Party membership on society generally. The political elite may have thought of the Party as the leadership cadre, the vanguard, of a coherent, rational program for Russian economic and political modernization. But large segments of the lower-class public appear to have thought of it as a means to improvement in personal social and economic status. Perhaps the political elite was guided by what they took to be sound theory rather than wishful thinking; my guess is that the public's perceptions were the product of direct observation of what was happening to their neighbors. In any event, the Party objectives—as laid down in Party documents in the mid-1920s—were not any better guides to the future development of the

---

67. See Azrael, *Managerial Power*, 12–27.

Party apparatus than Lenin's objectives as formulated in "State and Revolution" were for the state apparatus.

Finally, I would assert that the group of policies associated with the *vydvizhenie* at the very end of the 1920s were more "affirmative reaction" than affirmative action. I do not question that the behavior and the perceptions of Stalin, Molotov, Kuibyshev, and other members of the highest political elite were just those that Fitzpatrick and Bailes have ascribed to the leadership. I do wish to observe, however, that even if Stalin represented himself as fully in control—even if he thought himself fully in control—this does not negate the fact that he was responding to organizational and social developments that neither he nor anyone else specifically set out to achieve, results that were the aggregation of tens of thousands of choices made by thousands of individuals throughout the 1920s.

This study, beginning with the early nineteenth century, has observed with monotonous regularity that large organizations are hard to control. Large organizations whose structure and operational patterns are fairly clearly established, moreover, are exceptionally difficult to control from the top or from outside. This is so mainly because of the extended degree to which participants interact among themselves and with the surrounding society: most of these activities are beyond the capacity of the would-be controllers either to monitor or to guide. Such a complex system offers an ideal ground for misconceived and misguided efforts at control, especially in a traditionally authoritarian society. In such a social environment the ordinary reaction to the unanticipated consequences of any effort at control is simply to exert more control. In abstract terms, Jay Forrester, whose work on systems is unfortunately no longer fashionable, described the pitfalls in efforts to control highly complex systems intensively:

> In the complex system, when we look for a cause near in time and space to a symptom, we usually find what appears to be a plausible cause. But it is usually not the cause. The complex system presents apparent causes that are in fact coincident symptoms. The high degree of time correlation between variables in complex systems can lead us to make cause-and-effect associations between variables that are simply moving together as part of the total dynamic behavior of the system. Conditioned by our training in simple systems, we apply the same intuition to complex systems and are led into error. As a result we treat symptoms, not causes. The outcome lies between ineffective and detrimental.[68]

Trotsky's contempt and anger toward the new lower-class elites and sub-elites was understandable, given the complexity of the organiza-

---

68. Jay W. Forrester, *Urban Dynamics* (Cambridge, Mass., 1969), 9–10.

tions in which they worked. By the time he wrote his jeremiad against Stalin's revolution, some of the pre-elite of the 1920s were well on their way to elite status. It was they—not Trotsky and other old-school intellectuals—who, at the end of the revolutionary era, could be said to have won the revolution. But far from demonstrating Stalin's sole control of both Party and state administrations, the rise of the first Soviet pre-elite suggests instead that Stalin and his associates, like Trotsky, were reacting to circumstances beyond their control. These new elites and sub-elites, I propose, were more than the "ruling serfs," the bosses whom Lewin caricatures as having a "double face: the one looking down, of a despot; the other looking up, of a serf."[69] They were a force for self-interest that not only supported Stalin but urged him on to programs that could only be advantageous to their careers. It is not surprising that Trotsky, like many other intellectuals of the time and afterward, confused Stalin's successful machinations against the Oppositionists with total Stalinist control of Party and state.

69. Moshe Lewin, *The Making of the Soviet System: Essays in the Social History of Interwar Russia* (New York, 1985), 268.

CHAPTER 6

Bureaucratic Structure and Revolution: Elite Survival and Replacement

The problem of understanding any complex organization is daunting. Formal organizational rules, the application of technology to organizational tasks, and the politics and economics of mastering a wide variety of organizational objectives all contribute to the complexity. But, it is principally the relations among the people who are members of the organization or who must deal with it on some level that make the problem especially difficult. Crowding with them into their bureaus and waiting rooms are not only their previous experiences and their training, their capacity to get along, their intellectual and physical energy, and their personal ambition but the sum of their family histories, the demands imposed on them by their masters, and the expectations of society at large. Emerging out of their daily, seemingly pedestrian, actions are not only circumstances that affect how millions will live out their lives—or, sometimes, *whether* they will—but new and ever changing forms of power and status of which the bureaucrats themselves are active beneficiaries or victims.

In this sense, whether bribes are an issue or not, all bureaucrats and all political bureaucracies are venal—or, to put it another way, "corruption" is not a critical discriminator among bureaucratic systems. The Weberian ideal bureaucrat who is socially and psychologically disinterested in the outcomes his decisions will produce is no more than that, an ideal.[1] Historically, because of his or her membership in society and usually for other reasons in addition, the bureaucrat is either advantaged or damaged along with everyone else, both by what is decided and by the method of decision. And the very power to make such decisions, to apply experience, training, technique, and authority to areas that are reserved exclusively for his or her competence is inevitably a source of social and psychological recompense for the bureaucrat. Of course, such recompense is not sanctioned by law, and it is denied to those outside the system.

It is because disinterestedness is, on the whole illusory that the creation of a bureaucratic managerial class is equivalent, in modern society, to the creation of new social elites and sub-elites. Bureaucracies have become so large and their functions and functionaries so significant in determining social status that *their* elites are at least partly equivalent to *society's* elites.

But the task of this book is more complex than describing the most critical of these social relations at a given moment. I have tried to analyze how a really huge administrative body changed over a long period of time in Russia, a time and place where little else seemed so common as dramatic change. This concluding chapter attempts to summarize the findings of the study, to show how they relate to work that has been done by others on related problems, and to apply these observations to the broader landscapes of Soviet political development.

Individuals and Organizational Control

We have seen several illustrations of why the control of organizations by individuals should be treated as a major historical problem instead of as a conclusion foregone. For one thing, the organizations studied here all outlived their participants in a highly complicated, serial manner.[2] There were episodes of dramatic discontinuity, but most of the time participants came and went in relatively small numbers. This fact gives rise to the idea of an "organizational memory." Parts of the organizational structure—such as the way offices were organized and the way responsibilities were divided—may well have been the result of decisions taken long before the admission to office of any person serving

---

1. Max Weber, *Law in Economy and Society*, trans. and ed. Max Rheinstein and E. A. Shils (Cambridge, Mass., 1954), 220 ff. See also Reinhard Bendix's gloss on this subject in *Nation-Building and Citizenship*, rev. ed. (Berkeley, Calif., 1977), 141–42.

2. For a focused discussion, see Alvin W. Gouldner, "The Problem of Succession in Bureaucracy," in Gouldner, ed., *Studies in Leadership: Leadership and Democratic Action* (New York, 1950), 644–59. The problem as it bears on Soviet administration is discussed at length in Jerry F. Hough, *Soviet Leadership in Transition* (Washington, D.C., 1980).

at a given time. Such subdivisions of responsibility need not even have been written down; they may have been retained in the collective memory of participants—especially those who were advantaged by the arrangement. As Marshall Dimock observed more than a generation ago, institutions "seem to have a life of their own after they have been in existence for some time; it is as though they were the collective embodiment of all those devoted souls who have gone before.... The institution is stronger than the men."[3] At the very least, the evident problem of understanding how or whether organizations operate under human control compels us to conclude that their historical study requires techniques that are different from those we employ to write biography, the history of individuals.

The fact that organizational history is different in kind from individual history, like so much with which we have to deal here, is alien to our way of thinking about human social life. Such organizational characteristics as a memory that is clearly independent of any single kind of material resource, or of a specific individual, challenge our descriptive ability just as they prompt us to take seriously the kind of cartoon imagery that William McNeill uses: owing to the influence of modern bureaucratization and technology, he says, European armies "evolved very rapidly to the level of the higher animals by developing the equivalent of a central nervous system, capable of activating technologically differentiated claws and teeth."[4] Such language is at once attractive because of its comprehensiveness and repellent because of its manifest inaccuracy, but until we develop more adequate concepts and the terms to go with them, it will probably have to serve in many cases.

Perhaps the greatest source of confusion in the study of such organizations as we have been dealing with is that on the surface their operations and structure are apparently orderly, even logical, as that term is understood in classical psychology. In fact, both the assumption that there is a structural and functional logic in organizational behavior and the idea that organizational behavior can be explained as the sum of the behavior of the current organizational participants are misleading, but the difficulties engendered by the former misconception are rather more subtle. Our general inability, as rational actors, to anticipate the consequences of our individual actions is the essence of irony in literature. Moreover, it has been a circumstance of note for students of modern politics at least from Tocqueville to the present. "Most of all," Tocqueville wrote, "I am struck, not by the genius of those who served

3. Marshall E. Dimock, "Bureaucracy Self-Examined," in Robert K. Merton, Alsa P. Gray, Barbara Hockey, and Hanan C. Selvin, eds., Reader in Bureaucracy (New York, 1952), 397–406.
4. William H. McNeill, The Pursuit of Power (Chicago, 1982), 125.

the Revolution's cause, but by the stupidity of those who, without in the least wishing it, caused it to occur."[5]

More recently, commenting on the Revolution of 1917, the sociologist (and socialist revolutionary) Pitirim Sorokin wrote: "Counts and barons, landlords and business men all applauded scathing criticisms of the Government and acclaimed the approaching Revolution. These men, weary, effeminized, accustomed to lives of comfort, calling for revolution, presented a curious spectacle. Like heedless children, they manifested a curiosity and a joy in meeting such an interesting development. I had a vision of the French ruling classes before the eighteenth-century Revolution."[6]

But these are only examples of *individuals* who, in their enthusiasm, ignorance, or indolence, misjudged the outcome of an admittedly complex, short-lived phenomenon: social revolution. Their misjudgment of the logic of revolution may be understandable in the light of their limited experience and of the utter jumble of individual, group, and organizational actions and reactions associated with revolution. Bureaucracies, by contrast, are social phenomena with which we have daily and even intimate contact, yet we seem to misinterpret them too. One reason is that either as organizational participants or as historians of organization, we oversimplify the connection between cause and effect. In particular we tend to underestimate the organizational relevance of "feedback loops," the very common social circumstances in which effects become causes.[7]

More than this, we tend to ignore the ways in which cause-effect relations are affected by the accumulation of events, their history. As Forrester put it: "The present action stream corresponds to the present decision that in turn depends on the present information. However, the present level of the system does not depend on the present action but is instead an accumulation from all past actions. For example, consider a tank that is being filled with water. The height of the water is the system level. The level depends on the accumulation produced

---

5. Alexis de Tocqueville, *Oeuvres complètes: Edition definitive publiée sous la direction de J. P. Mayer*, 2:115, as quoted in Melvin Richter, "Tocqueville's Contributions to the Theory of Revolution," *Nomos* 8 (1967), 97. For additional examples, see ibid., 107, and Louis Gottschalk, "The Causes of Revolution," *American Journal of Sociology* 50, no. 1 (1944), 1–8.

6. Pitirim A. Sorokin, *Leaves from a Russian Diary—and Thirty Years After*, rev. ed. (Boston, 1950), 1–2; also Sorokin, *The Sociology of Revolution* (Philadelphia, 1925), 280–81.

7. D. K. Rowney, "How History Beats the System: Disaggregative Characteristics of Open Political Systems in Revolution," in A. J. Neal, ed., *Violence in Human and Animal Societies* (Chicago, 1976).

by the past flow of water, but the level is not determined by how fast water is being added at the present instant."[8]

Forrester might well have added, that the example does not reflect adequately the relation between cause and effect in most social organizations mainly because the example is too simple. Most feedback loops in social organizations are complex, requiring many iterations of input-output sets before a "result" can be identified. Similarly, most organizational "levels," or present conditions, are the accumulated product of many feedback relations.

Michel Crozier offers a more familiar description of the unpredictability of cause-effect relationships in bureaucratic organizations. Bureaucracy, he wrote, is an organization that "cannot correct its behavior by learning from its errors" because it always tends toward equilibrium among sources of conflict and dysfunction rather than toward what seems to be obscure and risky resolution.[9] Because of the degree to which sectoral equilibria fragment the bureaucracy as a processor of information or a transmitter of authority, it becomes difficult or impossible to exact uniform and predictable results from any set of inputs. Similarly, it is almost impossible to trace the many cause-effect relations that account for the consequences that can actually be identified.

It should come as no surprise, then, that either tsarist or Soviet political decisions on the composition of civil service elites, for example, should have proved impossible to implement. Both before and after the Revolution of 1917, elite organizational status was the congealed crust of a ripe stew of social, political, and economic ingredients, some of them internal to the civil service and barely controllable by it (such as educational qualifications for promotion); some of them external to the organization and, indeed, to the entire political system (such as the demand by certain groups for advanced civil service status). Skocpol is not alone when she exaggerates the centrality and dominance of the state apparatus as the "single strong point of the old regime, the keystone of the entire edifice." Indeed, I would say her error is more one of degree and emphasis than of outright misinterpretation. Other ele-

---

8. Jay W. Forrester, *Principles of Systems.* (Cambridge, Mass., 1968), 1:10.
9. Michel Crozier, *The Bureaucratic Phenomenon* (Chicago, 1964), 187. See also Robert K. Merton, "The Unanticipated Consequences of Purposive Social Action," *American Sociological Review* 1 (1936), 894–904; and "Bureaucratic Structure and Personality," *Social Forces* 18 (1945), 560–68; Alvin W. Gouldner, *Patterns of Industrial Bureaucracy* (Glencoe, Ill., 1954); Philip Selznik, *TVA and the Grass Roots* (Berkeley, Calif. 1949); McNeill, *Pursuit of Power*, 298–99, 60–61, 236–37, 277–79; Jean-Claude Thoenig, *L'ère des technocrates: Le cas des ponts et chaussées* (Paris, 1973); and for a summary of theory, Lucien Sfez, *Critique de la Decision* (Paris, 1981), 82–161, 212–58.

ments were critical, and many of these, as Skocpol shows, survived to exert their influence under the new regime.[10]

The attainment of organizational status within this apparatus was so crucial to so many segments of social, political, and economic life in Russia that simple political efforts at control through administrative law or political decree seem to have been foredoomed to failure. This is not an unfamiliar story. Because of ignorance, lack of sufficient ambition, or want of authority, reformers were not capable of attacking perceived organizational problems in all their complexity. As a consequence, the policies themselves, even in their earliest stages of formation—as Judge, Orlovsky, and Wortman show—became objects of social and political manipulation.[11] Examples abound of the failure of officials and politicians in Russia to anticipate correctly the outcomes of their actions.[12]

Commentators on such unexpected cause-effect outcomes as one encounters in Russian history are fond of the word "ironic." The true irony, it seems to me, is not that we derive so little from our most prodigious efforts to control complex organizations as to render the attempt Chaplinesque but that in the face of so many self-evident failures we continue to expect so much, either as organizational participants or as historians of organization.

A sounder historiographical sense of the limits of political control would have moderated our extreme interpretations of the role of politics both in tsarist and Soviet Russia. Even without fully understanding why, however, we have come to the conclusion that the idea of the omnicompetent or totalitarian state is inadequate to explain Russian or Soviet political history. The language used by Friedrich and Brzezinski or Schapiro when, twenty and thirty years ago, they described totalitarian politics now sounds remarkably dated.[13] Complex sociological, social-psychological, or social-political interpretations that are based on variants of the totalitarian model appear, on their face, to be possible

---

10. Theda Skocpol, "Old Regime Legacies and Communist Revolutions in Russia and China," *Social Forces* 55, no.2 (1976–77), 300.

11. Daniel T. Orlovsky, *The Limits of Reform: The Ministry of Internal Affairs in Imperial Russia, 1802-1881* (Cambridge, Mass., 1981); Richard Wortman, *Development of a Russian Legal Consciousness* (Chicago, 1976); Edward H. Judge, *Plehve: Repression and Reform in Imperial Russia, 1902–1904.* (Syracuse, N. Y., 1983).

12. See, e.g., George Yaney, *The Systematization of Russian Government* (Urbana, Ill., 1973).

13. Carl J. Friedrich and Zbigniew K. Brzezinski, *Totalitarian Dictatorship and Autocracy,* (Cambridge, Mass., 1956); Leonard Schapiro, *The Origin of the Communist Autocracy: Political Opposition in the Soviet State, First Phase, 1917–1922* (Cambridge, Mass., 1955). The question of whether the USSR was, or is, a totalitarian state continues to be debated in the Soviet studies community; see Stephen F. Cohen, *Rethinking the Soviet Experience* (New York, 1985), chap. 1, 2.

Bureaucratic Structure and Revolution 181

only in a social and political never-never land where organizations are merely passive, if somewhat clumsy, tools in the hands of political leadership. Take this remarkable 1954 formulation by Philip Selznik (better known as the author of *TVA and the Grassroots*), for example: "The bolshevik type of party is an effective organizational weapon because it has solved many of the problems associated with transforming a voluntary association into a managerial structure. This is the key to whatever mystery there may be about the organizational power of communism. Put most simply, the process referred to is one which changes *members* into *agents*, transforms those who merely give consent into those (at an extreme, soldiers) who do work as well as conform."[14]

For Barrington Moore, Jr., the state and its bureaucracy had "swallowed society" by 1954 in the USSR: "The behavior of nearly every adult male during his waking hours is heavily determined by his place within this bureaucracy," he wrote. Moreover, the entire system was so effectively interconnected that "the Party Presidium now constitutes the most important single device through which the actions of Soviet citizens are connected with one another."[15] More important, we are entitled to conclude, than money, kinship, and sex, even if the latter two are not exactly "devices" in Moore's sense.

The data examined in this study would not directly refute or confirm the visions of either Selznik or Moore of the role of the political bureaucracy in post–World War II Soviet society. But again and again the data do provide reason to believe that between 1800 and 1930 bureaucracy was not always a pliant tool—or a predictable one—in the hands of political leaders or of anyone else. My point is that the language in which these visions are couched seems appropriate only if we believe that Stalin or the Stalin system of the 1930s introduced an entirely unprecedented order of organization for bureaucracy, an iron logic and a kind of collective personality that were entirely unknown before in either Soviet or Imperial Russian history. Given both the conditions and the outcomes of bureaucratic development in Russia in the nineteenth and twentieth centuries, this seems doubtful at best—though not, I admit, entirely impossible. The burden of proof necessary to legitimate such language, nevertheless, inevitably falls on those who insist on its appropriateness.

But the same problems of language and concept crop up in efforts to criticize political leadership either for a lack of control or for an in-

14. Philip Selznik, *The Organizational Weapon: A Study of Bolshevik Strategy and Tactics* (Glencoe, Ill., 1960), 21.
15. Barrington Moore, Jr., *Terror and Progress—USSR: Some Sources of Change and Stability in the Soviet Dictatorship* (Cambridge, Mass., 1954), 2–3.

sufficiently nuanced control.[16] *Of course* political leadership is responsible for even the unanticipated outcomes of its initiatives. But in the absence of clear evidence that outcomes—alternative or otherwise—were clearly understood, or in the absence of the possibility of restructuring the "organizational weapon," of altering its social and political dynamics, of changing the environment in which decisions are both reached and implemented, the presumption should be against the notion that there were alternatives. This is an important point to which I return below.

To sum up then, the irrationality evident in unpredictable bureaucratic behavior is related to several factors. The public, political authorities and bureaucratic actors themselves are ignorant of the potential total range of their actions, and this ignorance combines—often destructively—with a tendency to assume that disinterestedness among bureaucrats is the norm for bureaucratic behavior. In fact, issues of status, professional competence, ego, public and private interest may always be counted on to exert their diverse attractions on any given action. Organizational behavior at any given moment, moreover, is the consequence not only of what someone may have done in the recent past but of actions taken perhaps years earlier, whichare, as a consequence, as much a part of the organizational environment as the office wallpaper and the pigeons on the window-sill. But the most significant factor impeding efforts either to control or to understand organizations would seem to be our general inability to project through all iterations the potential impact of any given action.

The Immobilization of Experts and the Maintenance of Elite Status

One of the most important details of organizational development since 1800 has nearly always been the status of experts. With the introduction of any technology more sophisticated than sword and saddle, pen and paper—indeed, with the introduction of any *new* technology—the problem of the introduction of technical experts necessarily follows. In this sense an "expert" is simply an official whose organizational status depends on the possession and systematic use of information and techniques that other officials do not have. In industry, warfare, and civil administration since the eigteenth century, therefore, there has never been any question in Europe whether experts would take part in administration; the flood of new technologies, which con-

---

16. Moshe Lewin, "The Immediate Background of Soviet Collectivization," *Soviet Studies* 17 (1966), 190–92.

stantly ran against intuition about how "work" was done or how power was managed, made experts an unquestioned necessity. The only real question was, under what conditions were experts going to be used? Would they be in charge of part or all of some operations? Or would they serve as acolytes to established generalists and organizational "pols and pros"?

For the Russian elites of nineteenth-century civil administration, as we have seen, this question was complicated, since their experts tended to come from (just slightly) outside the circle of traditional social elites of Russian society. The problem of the "conditions" of expert employment carried with it the question whether these *arrivistes* would be welcome to exercise not only technical but social and political functions—not only in elite organizations but in elite society as well. If so, if they were welcome socially, then sensitivity to their organizational status should correspondingly have diminished. If not, if the established or "natural" elites were always going to treat them as social inferiors, then their organizational subordination would have to be assured as well, for as Bennett has shown in theory and law and as we have seen in practice, the two statuses—social and organizational—were intimately joined.[17]

In the event, and in spite of many well-advertised exceptions, subordination—a kind of organizational immobilization of experts—was the prerevolutionary standard. This does not mean that they were systematically prevented from forming themselves into the kind of technically homogeneous corps that became so important to the status security of early French technocrats. As Armstrong notes, even from the early nineteenth century a kind of "attenuated corps" system developed in which technical, regulatory, and provincial specialists were grouped by education as well as by the location and substance of their career activities.[18] To these corps categories, in fact, Armstrong might have added physicians and public health specialists.[19]

Subordination of experts, nevertheless, was the standard. Even in agencies such as the Ministry of Ways of Communication, where *tekhnika* was everything—or should have been—experts were indeed finally on tap, not on top, even when they constituted virtually the entire managerial and executive staff. Those on top were the landed elites, or

17. H. A. Bennett, "The Evolution of the Meanings of *Chin*: An Introduction to the Russian Institution of Rank Ordering and Niche Assignment from the Time of Peter the Great's Table of Ranks to the Bolshevik Revolution," *California Slavic Studies* 10, no. 1 (1977), 1–43, and "Chiny, Ordena and Officialdom," in W. M. Pintner and D. K. Rowney, eds., *Russian Officialdom* (Chapel Hill, N. C., 1980), 162–89.
18. John A. Armstrong, *The European Administrative Elite* (Princeton, 1973), 223.
19. Roderick E. McGrew, *Russia and the Cholera, 1823–1832* (Madison, Wis., 1965), 32–40, 55–56, 125–27.

the bureaucratic elites without land or legitimate aristocratic pretension, but also without expertise.

Remarkably, even among the captains of incipient Russian industry, suspicion and hostility toward experts was endemic in the late nineteenth century. As Rieber notes, in affairs of urban politics during the 1870s and 1880s the Russian merchants "exhibited the same contempt for the Russian technical intelligentsia as that that had long pervaded hiring practices in their own factories."[20] The anti-expert bias, of course, was not unique to Russia. As McNeill notes, the post-crisis years after 1815 saw the reassertion of the rights of traditional privilege in Prussia and elsewhere.[21] Moreover, Armstrong has shown that Russia was by no means among the most backward of European countries in introducing the educational means to administrative advancement through expertise.[22]

With the rapid expansion of civil administration in the nineteenth century and, especially, with the addition of thousands of new offices requiring specific kinds of expertise, the historical problem in Russia is not how the number of exports expanded so rapidly but, why they remained subordinate to traditional social, political, and bureaucratic elites. Was it to the advantage of the autocracy that they should remain subordinate? This seems unlikely. What did the autocracy have to gain from the immobilization of expertise? What did it have to lose from the accession of experts even to the top administrative and political positions of the state? Or what did it gain from continued dominance by legal, military, and bureaucratic pseudoexperts? It does seem clear that the continued subordination of *arrivistes* experts was distinctly advantageous to the traditional elites, groups who consequently continued to dominate and who, in fact, expanded their presence in the state administration in the years before the Revolution of 1917.

This book cannot resolve the problem, but Blum's explanation for the lingering of the signs and behaviors of landlord-peasant relations long after emancipation in Europe adds to our understanding of the Russian case: "Just as the traditional superior status of the nobility persisted after the emancipation, so too did the social and political inferiority of the peasantry. The passage from the old order of legally sanctioned privilege to the open society in which economic role and not birth determined status was too sudden for all of its implications to be understood and accepted by those whose values had been formed by the old order. It was a moral and psychological revolution that

20. Alfred J. Rieber, *Merchants and Entrepreneurs in Imperial Russia* (Chapel Hill, 1982), 103.
21. McNeill, *Pursuit of Power*, 216–19.
22. Armstrong, *European Administrative Elite*, 223–27.

required a profound change in the social and mental climate that only time could bring."²³ Blum might more accurately have said that it was a moral and psychological change requiring a revolution for its completion. In any event, an explanation based on the proposition that profound social change is incomplete until the generation that has produced the change dies off is helpful but not fully adequate; the historical picture was far too complicated.

Blum notes further that, "the dilution of the preeminence of the nobility was accompanied and to a considerable extent caused by economic reverses suffered by many noblemen." True enough, so far as socioeconomic status is concerned. But so far as sociopolitical status is concerned, even if preeminence was diluted, it was by no means denied; until the end of the Old Regime the nobility—the *landed* nobility—continued to hold onto its preferments in local and central administration. Moreover, as we saw in Chapter 2, it was precisely the rate of decline in landed property that was associated with the strongest influx of the nobility into provincial state service.

In one sense, if an increased reliance by the gentry on state service seems perfectly reasonable—the gentry moving from privileged landed status to privileged administrative status as the viability of its agriculture declined—it is not necessarily the outcome that the historical literature predicts. Apart from Becker's work,²⁴ the Western literature leads us to expect not only a dispirited and economically shattered nobility but an impotent and socially contemptible one as well. This view is in sharp contrast to the findings reported in Chapter 2.

Nor is this interpretation unique to specialists in Russian history. The view that the nobility was terminally debilitated is critical to Skocpol's interpretation of the revolution in Russia: a want of noble strength in the countryside was the necessary precondition to successful rebellions among the Russian peasantry, which in turn were critical to the rise of the new regime.²⁵ I would modify this view by saying that rather than simple weakness, it was the link between landed nobility and noble officialdom that destroyed the influence of the state in the countryside. So comprehensive was the destruction and so deeply ingrained were the nobilitarian roles in 1917 that a full decade would elapse before—via collectivization—the Communist regime could forge a new, reliable system of rural control. Only in the 1930s, after the destruction

---

23. Jerome Blum, *The End of the Old Order in Rural Europe* (Princeton, 1978), 429–30.
24. Seymour Becker, *Nobility and Privilege in Late Imperial Russia* (Dekalb, Ill., 1985).
25. Theda Skocpol, *States and Social Revolutions* (New York, 1979), 128–40 and "France, Russia, China;" *Comparative Studies in Society and History* 18 (1976), 175–210.

of the traditional land commune and of independent peasant agriculture, would the massive fissure produced by the destruction of the provincial nobility in 1917 be repaired. Only then would the state be able to integrate the countryside reliably into the postrevolutionary system. I thus argue that because of the continued influence of traditional social elites, expert status throughout the course of the old regime in Russia remained subordinate at worst, ambiguous at best, by comparison to that of the landed and traditionally privileged classes. This status changed only with the disappearance of the old social elite.

The Status of Experts after 1917

The year of revolution, 1917, saw the beginning of the end of the generalist and of generalist administration in Russia. Lenin's image of an administrative world operating under the aegis of experts rather than either politicians or bureaucrats found its expression both in the emergence of experts at the top levels of administration in central government and also in the restructuring of that government.[26]

From 1917 forward, the imperial generalist administrations were broken up, subdivided into operationally specialized commissariats. At the same time, generalist bureaucrats—whose organizational and social status in the old regime had been so intimately interconnected—were displaced, often by their former expert subordinates. Experts emerged, for almost the first time, both at the highest levels of civil administration and in the highest administrative councils, such as VSNKh and the Council of People's Commissars. To assert, as Bendix does, that intensive use of experts and of scientific management were a kind of substitute for the two hundred years of moral and religious formation that preceded industrialization in the West dramatically understates the complexity of a case that appears to have responded to the availability of experts and the departure of generalists, as well as to the political, social, and intellectual ambitions of the political leadership.[27]

To be sure, the initial efflorescence of technocratic ambition was a false start, or not really a start at all. During World War I, under the press of ever more demanding exigencies, the role of experts in transport, heavy industry, and munitions manufacture had waxed stronger, although not with notable positive effect on the war effort.[28] Since one thing the revolution most assuredly had not brought was peace, exi-

---

26. V. I. Lenin, *PSS*, 5th ed. (Moscow, 1963), 42:156. See also Eric Olin Wright, "To Control or To Smash Bureaucracy: Weber and Lenin on Politics, the State, and Bureaucracy," *Berkeley Journal of Sociology* 18–19 (1973–75), 88–89.
27. Bendix, *Nation-Building*, 197–208.
28. Armstrong, *European Administrative Elite*, 289-90.

gencies—whether the product of military disaster or of famine and disease—continued, and so did the demand for expertise. But the educational and organizational-industrial debates of the 1920s could leave little doubt that the false start was being seconded by a firmer and more decisive advance, one that underscored the importance of expertise and specialization, enhancing the organizational roles of economists, engineers, and scientists in the decade following the revolution.[29]

Contrary to Lenin's hopes and predictions, however, the advent of experts and specialists into the highest reaches of Soviet civil administration did not bring in its wake the demise either of politics or of bureaucracy. Party and civil government roles were from the very first so closely intertwined that the exclusion of politics and politicians would have been difficult at best. As time passed, debureaucratization and depoliticization were systematically reversed by the aggressive insistence that Party members should have not only oversight of official appointments but the benefit of technical education *and* preferment for technical appointments—including, of course, those at the very top.[30] Both in choice and in language, moreover, as Lewin notes, preference for simple, direct action (results! above all, results!) often overshadowed both the personal and professional interests of specialists and experts.[31]

Thus, the peculiar brand of Soviet expert administration began to emerge: vastly expanded technical-professional elites found themselves in charge of ever increasing spheres of economic and domestic administration in a system of intense political bureaucratization and centralization. For the post-1929 generation, as a mass of studies has testified, this meant rapidly expanding opportunities in economic, technical, and scientific administration—not only in education and civil administration but also, of course, in industrial administration and the Party. Engineers, economists, and agronomists—whatever Stalin said or felt or tried to do about them—came to dominate, by both their numbers and their roles, not only in civil and industrial administration but even in the Party. Whatever else may be true about the changed status of experts after 1917, the pre-1917 era of expert immobilization was clearly over.[32]

---

29. Sheila Fitzpatrick, *Education and Social Mobility in the Soviet Union* (Cambridge, 1979); Moshe Lewin, *Political Undercurrents in Soviet Economic Debates* (Princeton, 1974), 73–96.
30. Kendall E. Bailes, *Technology and Society under Lenin and Stalin* (Princeton, N. J., 1978), 159–87; Fitzpatrick, *Education and Social Mobility*, 113–35, 180–205.
31. Lewin, *Political Undercurrents*, 100–101.
32. For the background and career characteristics of both technical and political-technical cadres, see Bailes, *Technnology and Society*, 431–41. Educational and specialization data on pre-1959 appointees to civil, military, and political positions are found in Grey Hodnett, *Leadership in the Soviet National Republics* (Oakville, Ont., 1978),

That the emergent technical cadres were a new "Soviet bureaucratic elite" is by no means universally agreed. Armstrong, for example, applies such terminology to the Party apparatus rather than to civil or economic administration.[33] For Soviets, of course, the organizational homes of the emergent cadres were not bureaucracies at all but administrations or apparatuses.

According to Alex Simirenko, moreover, the term "elite" is itself a misnomer. This cadre formation was a process in which not only the specialists but the entire bureaucratic apparatus was both rendered subordinate to and corrupted by the "professionalization" of Soviet politics. On balance, the rise of the expert to senior managerial status in Soviet society is not a net gain, in this last view: the emergence of expert professionals is a creation of horizontally segmented groups more than of vertically distinct elites; moreover, not only elite status but even their ability to follow their professions is denied the newly emergent experts by the professional politicians of the Party. "In countries with professionalized politics other professions suffer," Simirenko writes, "since they cannot be permitted the autonomy and the uninterrupted routine found in the rational order. One vocation whose professionalization is particularly affected by professionalized politics is that of bureaucracy, since its professionalization is built on the principle of value neutrality."[34] Celebrated cases in which professional development and intellectual freedom appear to have been corrupted or distorted seem to give substance this view, even though it unrealistically assumes the existence of "value neutrality."[35]

In Odom's view, by contrast, the question is not so much one of professional independence as whether all segments of the administrative apparatus are uniformly subordinated to the political leadership. In his study of a volunteer organization with a heavy professional-technical emphasis (the Society of Friends of Defense and Aviation-Chemical Construction, or *Osoaviakhim*), he finds that the capacity of the senior political leadership to manipulate either the nonprofessional membership (the "masses") or its professional administrators (the *apparat*) was limited. "Drawing up rules for Osoaviakhim and inviting mass participation did not inspire mass compliance," he writes.[36] Odom details several ways, moreover, in which the members of the admin-

---

Tables 4.3–4.6. See also Jerry F. Hough, *The Soviet Prefects* (Cambridge, Mass., 1969), chap. 3, esp. Tables 2–4.

33. John A. Armstrong, *The Soviet Bureaucratic Elite: A Case Study of the Ukrainian Apparatus*. (New York, 1959).

34. Alex Simerenko, *Professionalization of Soviet Society*, ed. by C. A. Kern-Simirenko (New Brunswick, N. J., 1982), 12–13; also 11, 31–32, 42ff.

35. David Joravsky, *The Lysenko Affair* (Cambridge Mass., 1970).

36. William E. Odom, *The Soviet Volunteers: Modernization and Bureaucracy in a Public Mass Organization* (Princeton, N. J. 1973), 271.

istrative apparatus systematically evaded or manipulated central authority to their own personal, local, or career advantage.[37]

Although Odom's study is unique in its comprehensiveness and elegance, few of his findings are novel so far as our understanding of large organizations generally is concerned. Similar tales abound, on both an impressionistic and an empirical level, in both Soviet and Western studies of Soviet industrial administration.[38] But for stories of this kind the Soviet satirical newspaper *Krokodil* would probably go out of existence; it would certainly be deprived of the largest part of its humor, just as the Central Committee's weekly, *Ekonomicheskaia gazeta*, would be deprived of most of the material for its cartoons.

To say that administrative units at various levels of the bureaucracy tend to find ways of maintaining their independence from the political center does not demonstrate that professional and technical managers dominate Soviet civil administration in the way that we expect a technocratic elite to do.[39] It only supports the notion that political elites—as distinct from managers and experts—are not necessarily the dominant and controlling factors in Soviet administrative life today that we sometimes assume they are. There may well be scope for both personal and professional values in the Soviet administrative system.

On the other hand, the two views of administrative-political relations sketched above do demonstrate that serious scholarship is by no means in accord concerning the relative strength of the Soviet technical-managerial elite after 1929. To examine this relationship in detail would go considerably beyond the original objectives of this book. Suffice it to say, the foundations that Party and civil officials laid in the 1920s certainly changed the status and career expectations of administrative specialists, professionals, and experts as compared to those of the pre-1917 era. At the same time, they appear to have fallen short of producing the technocratic state system, envisioned by Lenin, in which politics and bureaucracy are subordinated to the rule of *tekhnika* and the technically optimal solutions to managerial problem are both clear and uniformly acceptable to all participants.

Status Maintenance and Organizational Adaptation

One of the most striking phenomena we have seen is the capacity of certain groups within the service elite and upper-class society more generally to adapt to changing organizational and political circumstan-

---
37. Ibid., 281–91.
38. David Granick, *The Red Executive: A Study of the Organization Man in Russian Industry* (New York, 1960); Hough, *Soviet Prefects*, 289–316.
39. Thoenig, *L'ère des technocrates*, 143–91, 231–51.

ces. Although specialist roles were created in large numbers in the nineteenth century and although these roles were assumed by socially distinct groups, it is evident that the social classes more traditionally aligned with bureaucratic office holding adapted to changing social and organizational circumstances, acquiring the skills necessary to manage and subordinate the experts. Again in the 1920s we found adaptability in the capacity of the prerevolutionary sub-elites to serve in large numbers at administrative levels and as specialists in the Soviet administrations. Although the acceptance of Party membership was not as common among the Soviet specialists—the professional sub-elite of the early 1920s—as among other groups, various other forms of adaptation were evident. These included acceptance of rapid organizational change, change of policy, and change of leadership.

Of course, any evidence of adaptability among bureaucrats flies in the face of at least one major theme in the literature on the history of bureaucracy and among popular notions of what constitutes the "bureaucratic" mentality. When we do not imagine them to be mindless instruments in the hands of political masters, we suppose that bureaucrats and bureaucracies are impermeable to structural change introduced from without, especially if the change is not evidently in the best interests of either the organization or of its personnel. This observation, as we have seen, is quite possibly founded on the fact that it is very hard to predict the outcome of a particular organizational change even under the best of circumstances. Our failure to understand, or organizational leadership's failure to predict, an outcome correctly is interpreted as organizational intransigence. In the nineteenth century, however, it seems evident that failure to change and to adapt constructively would have carried with it the certain decline of the nobility as a social group even before 1917. Yet the capacity to adapt, to shift from landowning to bureaucratic elite, from military to civil elite, does not provide quite the evidence we expect of either intransigence or weakness among traditional elites—features that are supposed to be targets for destruction in social revolutions.[40]

Lawrence and Jeanne Stone have shown that whether revolution is the outcome or not, elite adaptability can take highly varied, inventive forms across long periods of time. Of the six factors that the Stones identify as accounting for "elite stability," at least two—commercial foreign policy and successful family strategy—require a light-footed maneuverability and quickness of wit not normally associated with

---

40. Lawrence Stone, "Theories of Revolution: The Third Generation," *World Politics* 18, no. 3 (1966) 166; Gottschalk, "Causes of Revolution," 7; Richter, "Tocqueville's Contributions," 103.

sated and besotted upper classes. A third factor, cultural cohesion with the "middling sort," requires an openness to social change on the part of both aristocrats and commoners that seems rather unlike the Russian socioeconomic exclusiveness of the nineteenth and twentieth centuries. In the British case, the Stones point out, "this middling sort," rather than resenting their social superiors, "eagerly sought to imitate them, aspiring to gentility by copying the education, manners, and behavior of the gentry.... However it was achieved, the fact remains that the great strength of the English landed elite was its success in psychologically coopting those below them into the status hierarchy of gentility."[41]

One kind of openness, however, does not guarantee other kinds. For example, career adaptation does not presuppose cohesion with other social classes. Certainly one would not easily find among the characterizations of the Russian nobility and their relations with inferior classes the same kind of social and cultural flexibility that Stone and Stone associate with elite survival in England. Rieber, for example, describes as a kind of fixed circumstance of Russian social and political life the rivalry, lack of cooperation, and even contempt that poisoned relations between the merchant class and the nobility.[42]

Indeed, there is some suspicion that it was precisely *because* of their access to alternative, nonlanded status, a status that carried with it continued guarantees of legal privilege, that Continental elites maintained their great distance from the non-noble. "In England," Stone and Stone note, "the key concept of gentility was more dilute and ill-defined than on the Continent."[43] To be sure, the Stones link this phenomenon to the ability of would-be elites on the Continent to purchase offices and thereby become noble, a procedure that had its equivalent in the Russian scheme of things even in the nineteenth century. This condition gave rise, they argue, to careful distinctions among grades or vintages of nobililty that were simply nonexistent in England. Evidently, we need to see the flight to officialdom that characterizes the career history of the Russian nobility in the nineteenth century as more than an adaptation to a new source of income. After emancipation it was clearly a way of guaranteeing continued de facto as well as de jure noble status.

The capacity of Russian landed elites in the nineteenth and early twentieth centuries to maintain and indeed expand their hold on access to education and preferred offices remains one of the most eloquent

41. Lawrence Stone and Jeanne C. Fawtier Stone, *An Open Elite? England, 1540–1880* (Oxford, 1984), 407–21.
42. Rieber, *Merchants and Entrepreneurs*, 92–103, 136–37, 426–27.
43. Stone and Stone, *Open Elite?* 408.

illustrations of the complexities of interpreting the relation between social and administrative histories. It does seem reasonable to believe that as their hold on landed resources declined, the Russian nobility would be submerged by a rising tide of non-noble classes both in agriculture and in administration. It also seems reasonable that the autocracy should have been able to orchestrate the modernization of its administration without reference to the social interests of these decaying traditional elites—allowing non-noble, nonlanded officials to find their own level throughout the civil service, to their own advantage and to the advantage of the service and the autocracy. After all, what did "centralizing Tsars determined to extract from the Russian people sufficient resources to support military forces for defense and expansion in threatening geopolitical environments" (as Skocpol discribes them) have to gain from propping up the landed nobility in jobs that were the equivalent of make-work welfare for nineteenth-century elites?[44] "Sandwiched between the mildly commercialized serf economy and the Imperial state was the Russian landed nobility. Like the French proprietary upper class and the Chinese gentry, this Russian dominant class appropriated surpluses both directly from the peasantry and indirectly through remuneration for services to the state. But in sharp contrast to the French and Chinese dominant classes, the Russian landed nobility was economically weak and politically dependent vis-à-vis Imperial authorities."[45]

But social and organizational status in the nineteenth century, as in the twentieth, was so closely bound together that the upper classes' motivation for retaining both is clear. What is also clear is that the Russian landed elites not only had the capacity and the motivation to maintain their hold on the upper reaches of the civil service but that they did this successfully, tightening their grasp right up to the end of the tsarist era. Bureaucratization and centralization in Russia, far from demonstrating the weakness of the landed elites, seem to underscore the opposite—their ability to adapt and to use their diminished landed and legal resources to their own advantage even when it flew in the face of logic and even when it was to no one else's advantage. In the end, indeed, it was the death grip of the landed elites on the civil service, and especially on the provincial service, that brought that part of the Old Regime down with such a crash of suddenness in 1917. In short, it was their enduring strength, not the opposite, that resulted in their being engulfed in 1917, just as Tocqueville, Gottschalk, and the Stones lead us to believe it should have done.

44. Skocpol, "Old Regime Legacies," 292.
45. Skocpol, *States and Social Revolutions*, 85.

The Upward Mobility of Lower Classes

The data show that social classes from beyond the pale of traditional elites were inducted into state service in Russia long before the Revolution of 1917. Thus, for these social groups as for prerevolutionary experts, the relevant question is not whether the upward mobility of lower classes occurred but under what conditions it occurred.

Throughout the nineteenth century and, in fact, well before, the higher civil service and even the civil elite made way for a limited number of sons of the lower classes.[46] But the systematic accession of lower classes to all types of managerial positions and to elite status, is a postrevolutionary phenomenon. The reason, I argue, was not the weakness of noble elites but, on the contrary, their ability to hang onto high official positions and to the status that these positions guaranteed. After 1917 the process of modifying the terms of landholding, restructuring the social system, and reassigning incumbencies in the higher civil service each entailed the displacement of the Russian landed nobility. In other words, these rather rudimentary revolutionary objectives made either the abolition of all elites or the creation of an entirely new elite a foregone conclusion.

In the event, whatever the original intentions of the revolutionary leadership, what actually seems to have occurred is that a new elite was formed. To say this, however, is not at all to say that the social objectives of the revolution went entirely unfulfilled, although such a view (or variants of it) is common. Robert Conquest, for example, writes:

> Critics of Lenin are entitled to say that in the long run his principle of the centralized party and its vanguard role meant that his revolution could no longer be described in the Marxist categories, and amounted to substituting for the old regime the rule not of the working class but of a political and bureaucratic elite.
>
> In spite of his constant complaints about bureaucracy in his last years, no such possibility seems seriously to have entered his head. The Bolshevik grip on power was, he seems to have thought, the only and necessary precondition for realizing his more general ideas. With this once accepted, the complete alienation of the proletariat could be interpreted as a proletarian dictatorship, the tighter and increasingly autocratic centralism as a higher form of democracy.[47]

---

46. Borivoij Plavsic, "Seventeenth Century Chanceries and Their Staffs," in Pintner and Rowney, *Russian Officialdom*, 19–45; and Brenda Meehan-Waters, "Social and Career Characteristics of the Administrative Elite, 1689–1761," in ibid., 80–84.

47. Robert Conquest, *V. I. Lenin* (New York, 1972), 127.

Nor is Conquest's view in this matter at all exceptional. Sorokin wrote of the reconstruction, from the early 1920s, of the pyramid of power and status.[48] Ferro interprets the process of postrevolutionary bureaucratization as the creation of a new bureaucratic class, described in terms that make one think more of caste than of class.[49] In sum, the view that what was created was some species of new bureaucratic social elite is too common either to be ignored or entirely denied.[50]

But as we have seen, the historical process of building or rebuilding the Soviet bureaucracy cannot be fairly described as "the complete alienation of the proletariat." On the contrary, it relied heavily on the children of the proletariat and the peasantry and on active peasants and workers themselves to staff its thousands of offices, even at the enormous and possibly deadly cost of sacrificing education and expertise.

At the same time, the idea that this administrative "machine" was created to serve the needs of the Party exclusively and, especially, that it was seen as playing this role is impossible to sustain either from the sociodemographic data or from the recorded views of Bolshevik leaders such as Lenin, Trotsky before his exile, and Bukharin. Both the postrevolutionary Party apparatus and the new civil bureaucracy, whatever other roles they filled in Soviet society, were distinctly means for rapid upward movement of tens of thousands of workers and peasants throughout the 1920s. If this structure was an elite, it was the first administrative elite in European history to have been drawn in substantial numbers from the lowest social classes. Moreover, to the extent that it was proletarian and peasant, the new Soviet administration was impaired in its capacity to play the very role that its critics assign to it—the role of mindless, mechanical instrument of Bolshevik-Stalinist political power. Throughout the 1920s and probably the 1930s as well, the civil administration had at its disposal nothing like the education, expertise, and organizational experience necessary to fulfill this function, even if we assume that the will to do so was there.

A further complication to the interpretation of the upward mobility of lower classes after 1917 is the fact that there were many offices in which experience and expertise were indispensable and which, therefore, could not be filled by the upwardly mobile lower classes but were staffed instead by former employees of the tsarist bureaucracy. This meant that for at least ten years following the Revolution of 1917 thou-

48. Pitirim A. Sorokin, *The Sociology of Revolution*, (Philadelphia, 1925), 258.
49. Marc Ferro, *Des Soviets au commuism bureaucratique* (Paris, 1980), 110, 125ff.
50. Charles Bettelheim, *Class Struggles in the USSR: First Period, 1917–1923* (New York, 1976), esp. 271–74, 333–344; Roy Medvedev, *On Socialist Democracy* (New York, 1975), 127.

sands of important positions in the middle and upper reaches of both the All Union and Union Republic administrations, as well as a good many *oblast*-level offices, were filled by individuals in their forties and fifties who had received their education and early career experience under the last tsar.

Needless to say, these were neither workers nor peasants; their education was far too good. Under the Old Regime, members of the lowest social classes would not have had such educational opportunities. But neither had these persons been members of the prerevolutionary social or administrative elites. They were *sluzhashchie*, that curious creation of the Soviet political mind—not a true social class but an employment category that carried with it the assurance that the holder of the status was, de jure, free of antirevolutionary taint. The prerevolutionary positions they had held tended to place them not only out of the ranks of the elite but even below the ranks of the higher civil service. For this reason I speculate that these newly emergent Soviet technical elites must have welcomed the opportunity that the revolution presented even when they chose not to become Bolsheviks or to support the proletarian cause in any way other than by doing their jobs.

The *sluzhashchie*, or white-collar workers, succeeded in holding down thousands of jobs that were no doubt the envy of the worker-peasant Party members who moved into civil administration with them during the 1920s. Eventually, of course, the first generation of *sluzhashchie* gave way. For what must have been a combination of actuarial and political reasons, by the mid-1930s the positions held by the old *sluzhashchie*—not only in state and Party apparatuses but, as Nicholas Lampert shows, in industry—were added to the host of vacancies opened up by the rapidly expanding economic and industrial administrations.[51]

Intergenerational Change in the Revolutionary Period

Considering that the role of any given organizational generation in shaping the future of the organization is necessarily large, I have paid relatively little attention to the problem of intergenerational change. In part this is simply because most ordinary features of such change are themselves unchanging; that is, organizational generations tend to produce successor generations in their own image and likeness. Such replication reinforces the rule of patriarchy just as it tends to select out of

---

51. Nicholas Lampert, *The Technical Intelligentsia and the Soviet State* (New York, 1979), chaps. 3, 4.

the general population certain social and ethnic groups that are attractive to the organizational elite. This circumstance is a source of frustration to some would-be participants, just as it is to students of organization. Armstrong, for example, is constrained to comment on the lack of interesting findings in the wake of a painstaking comparison of career development and office attainment of major European administrative elites in the course of several decades.[52]

Whatever the normal pattern of transfer of authority and responsibility from one generation to another, it is clear that revolution has a high potential for disrupting it. In Russia it is evident that a return to "normal" patterns was delayed even beyond the departure of the revolutionary generation itself. As a conclusion to my analysis, therefore, a consideration of these rhythms of change is in order.

Between 1890 and World War II one may distinguish perhaps three generations of Russo-Soviet administrators: the prerevolutionary generation; the first postrevolutionary group, dominating the scene up to 1928; and the second postrevolutionary group, entering during the 1920s and becoming dominant during the 1930s.[53] It is important to note that the generation of 1890–1917 is the only one of these three most of whose members entered service and were promoted by what we could call normal processes of turnover. Tsarist turnover patterns varied relatively little from decade to decade; indeed, career development in broad terms was similar from one major European civil service to another.[54] The way in which such stable patterns of career inauguration and development shape both careers and organizations is important, of course. But for the moment I am more interested in the effect of the latter two "atypical" generations.

The first of these was the one to which the revolution gave power. The force of revolution exposed this generation to extreme disruption, such as the rapid turnover of the 1917–18 era, an unprecedented change from the past. One may suppose that many of these administrators were children of the upper classes; however, they no longer reflected the noble domination of Russian politics and administration that was common before 1917. The upper levels of the administration, moreover, consisted of people who had never been prepared for their offices in what would have been regarded, before 1917, as the normal way. Whether they were Bolsheviks or not, the new administrative elite seems certainly to have consisted of people who, five years earlier,

52. Armstrong, *European Administrative Elite*, 241.
53. Interpretations vary; eg., Hough, *Soviet Leadership*, 37–60, identifies three Soviet generations through World War II. See also Seweryn Bialer, *Stalin's Successors: Leadership, Stability, and Change in the Soviet Union*, (New York, 1980), 63–126.
54. Armstrong, *European Administrative Elite*, 239–41.

could not have expected to hold senior government positions. Moreover, because at accession they were middle-aged, and because of political factors (discussed briefly below), a considerable portion of this generation had ceased to be active by the middle 1930s. They were displaced, that is, by a revision of education and promotion standards, by ordinary actuarial factors, and, more directly, by the purges of the apparatus of the Party and the state administrations in the 1920s and 1930s.

Surely the most obvious fact about turnover between 1917 and the early 1920s is that of variation in social class, whatever the profundity of that change. The revolution witnessed the collective and definitive dismissal of the nobility—especially the landed nobility—from their six-hundred-year dominance of the upper levels of state service. Whatever may be concealed within the terms "employee," "white-collar worker," or "sluzhashchie" in the data from the 1920s, it is clear that the class category that dominates the data before 1917—the hereditary nobility, or dvorianstvo—has simply disappeared from view, probably forever.

Since such a high proportion of top offices was held by nobility before 1917, and since the relationship between career seniority and top officeholding is clear, we may infer that normal patterns of relations between senior and junior officials were disrupted. But if important consequences flowed from the disruption of the traditional pattern of dominance of juniors by seniors, even more significant consequences flowed directly from the disappearance of the dvorianstvo. For example, the class-selective or "closed" educational institutions that had arisen in the eighteenth and nineteenth centuries were overthrown. The preparation of elite civil servants, which had been the responsibility of the Tsarskoe Selo Lyceum (Alexander Lyceum) and the Imperial School for Legal Studies, was simply abolished as a function, and the role of those schools as training grounds for administrative leaders was not entirely replaced for many years.[55] Elites as they were known and cultivated in the nineteenth century ceased to exist, and elitism took a different form, as we have seen.

Of obvious importance is the impact of the class or complex of classes that emerged to take the place of the noble officials. To be sure, the social group sluzhashchie was enormously important in securing continuity in administrative function. Inevitably, however, its role was temporary. Even if its members had not been perceived as "bourgeois specialists," aliens temporarily permitted to serve the new order, their

---

55. Hough, Soviet Leadership, 118–19; Jeremy Azrael, Managerial Power and Soviet Politics (Cambridge, Mass., 1966), 157–62.

disappearance would be predictable on actuarial grounds. As a group they varied in age, but one can assume that most members of the cohort educated before the Revolution of 1917 were, by the end of the 1920s, at least in their forties, and the data have shown that many of the holdovers who managed to garner high-level positions were considerably older. Some in the group, then, could be useful for perhaps another twenty years, but most would pass from the scene within a decade. They would perforce disappear under the wave of a new Soviet generation, no matter what the political environment.

But such projections into the future would have assumed a stable administration, a stable economy, a stable population, and a stable political environment—none of which was present in the 1920s. While deliberate efforts were being made in Moscow to limit the size of the existing administration, problems of economic development, the repopulation of cities ravaged by civil war, riot, disease and famine, and the political demands of Communists whose status did not yet conform to their ambitions must all have combined to call into question the serviceability of the existing administrative elite. In other words, it may well have been the case that a Stalin or a Molotov wanted to achieve a rapid turnover that would put Communist workers in positions of influence, but it may also have been the case that the young subordinate ex-workers and ex-soldiers in state administration and the ambitious young Communists in industry were pressing intensely for their own upward mobility.[56] Even without these pressures it is evident from the data that major discontinuities in the population of state servitors would have caused significant problems for the simple availability of expertise both in the 1920s and afterward.

Thus, the 1920s appear not merely as a moment of instability and ferment in the history of Soviet administration; taken as a whole the decade was much more. It witnessed the introduction of the first specialist commissariats and the first industrial management commissariats. From the point of view of structure alone, these years witnessed the beginning of government dominated by specialists and professionally specialized institutions, a transition to technocracy. The period also witnessed changes and significant ruptures in the pattern of personnel turnover the like of which had not been seen in Russia in modern times. This said, however, one must note that such changes were by no means a straightforward replacement of tsarist elites. The complexity

56. For an extended discussion of the relation between upward social mobility in the USSR and its impact on our understanding of the Revolution of 1917, see Sheila Fitzpatrick, "The Russian Revolution and Social Mobility: A Re-examination of the Question of Social Support for the Soviet Regime in the 1920s and 1930s," *Politics and Society* 13, no. 2 (1984), 119–41.

of the replacement process reflected the complexity of circumstance in which officials found themselves, not to mention the audacity of the goals that politicians had set them.

The replacement process worked against *social* elites, it is true, but not uniformly against prerevolutionary organizational elites and sub-elites. Many of these survived to occupy secure organizational positions at the elite and sub-elite levels in the 1920s. To an important extent the successful survival of the prerevolutionary cohort of administrators was a result of the complex character of the preparation and experience of the prerevolutionary elites themselves. In a way, that is, they were in a good position to ensure their own survival whether there was a revolution or not. The Revolution of 1917 simply enhanced that postion.

One way of understanding this influence is to suppose that the nineteenth-century industrialization and urbanization process in Russia had less intimately involved the state—as it did, for example (in the illustration offered by Eisenstadt), in England, Holland, and the Scandinavian countries. The result would have been a slower bureaucratic transition within a smaller scope of bureaucratic activity. In a few Western countries, according to Eisenstadt, "this smaller scope was due both to the high extent of self-regulation and political articulation of the principal strata ... and to their great influence on the goals of the polity. The fact that their bureaucracies did not undergo processes of aristocratization and alienation explained many aspects of the type of gradual political change to which these countries were subject."[57] In Russia, of course, this would have meant that fewer experts or specialists of any kind would have had government experience or the ability to function in the environment of a state bureaucracy.

Such a reduced level of state interventionism, in turn, might have forced the modernizing political leaders of 1917 to rethink the means they were choosing to achieve social and economic change, since they would have confronted the expert cadres on entirely different terms. Surely a less vigorous transitional administration would have slowed the creation of specialist commissariats, just as it would have reduced dramatically the specialist elites and sub-elites who were available in 1920 for state service.

Nor would the lower-class *arrivistes* of the 1920s or the upwardly mobile *vydvizhentsy*—the principal products of the Cultural Revolution of 1928–32—have had the same impact: not only would the management and executive cadres who were there to shape the upcoming elites have been absent, but so would the institutions that employed

57. S. N. Eisenstadt, *The Political Systems of Empires* (New York, 1963), 356.

them. Whatever one may think of the New Soviet Man, his formation, his ambitions, it is important to remember that he was the grandson of those who lived and worked in the Russia of Alexander III. Whatever the degree of social or organizational discontinuity, the solutions to the political, economic, and institutional problems of the 1930s were, by this same logic, shaped in the framework of patterns of social, political, and economic behavior of the nineteenth century.

Reinhard Bendix and J. L. Talmon explain the political universalism—or what Bendix calls "plebiscitarian domination"—of Soviet society by writing of the revolution's "debt" to the French Revolution and its tendency to annihilate politically mediating groups or structures, the Toquevillian *pouvoirs intermédiaires*.[58] Bendix, indeed, goes somewhat further: "The story of Western models and Russian responses can best be told consecutively, for the gradual transformation of Russia's "backwardness" is indistinguishable from its response to the West."[59] But it is vital to give due recognition to the fact that in Russia such power had been annihilated long before either the Revolution of 1917 or even that of 1789. Although Bendix does appear to recognize the traditional monopolistic tendencies of the Russian state, he is eager to return to the Westernizing explanation as fundamental to the state's political development.[60]

Of course, even on my view it is possible to go too far. Perry Anderson's notion that the Russian state was "the only absolutist state to survive intact into the 20th century" is, at best, misleading even if we note his qualification that the "social formation was dominated by the capitalist mode of production."[61] Bureaucratization and the centralization of political authority had moved steadily forward throughout the nineteenth century. Rather than parts of an absolutism that was about to disappear, they should be understood as the materials out of which a new Russo-Soviet state was being fabricated. I thus conclude not only that the roots of Soviet plebiscitarianism are not found in the French Revolution but, equally, that it vastly overstates the case to assert, as Anderson does, that the "Russian Revolution was not made against a capitalist state at all. The Tsarism which fell in 1917 was a feudal apparatus: the Provisional Government never had time to replace it with a new or stable bourgeois apparatus."[62]

---

58. J. L. Talmon, *The Origins of Totalitarian Democracy* (New York, 1960); Bendix, *Nation-Building*, 175–77.
59. Reinhard Bendix, *Kings or Peoples: Power and the Mandate to Rule* (Berkeley, Calif., 1978), 493.
60. Bendix, *Nation-Building*, 179.
61. Perry Anderson, *Lineages of the Absolutist State* (London, 1974), 353, 359–60.
62. Ibid., 359.

The Revolution of 1917 neither destroyed an absolutism nor inherited one. It is clear, however, that many of the traditional structures and rules of the game of governing were potent and that the revolutionary leadership carried them across the divide of 1917. One of these—a "rule" that manifests vital aspects of the authority structure of the nineteenth century—was that induction and advancement in service should be based on the class principle. A preoccupation with social origins and a profound commitment to the notion that organizational effectiveness and political loyalty were dependent on class background were no more alien to the Russia of Lenin than they were to the Russia of Alexander III or Karamzin. It is equally clear, however, that included in the baggage of hand-me-downs were other operative values, other assumptions about how to get things done and about what things needed to be done first and by whom.

The realization that nineteenth-century administrative structures were a force in 1917 or in the 1920s should not be taken to render vacant the debates of the 1920s about what alternatives were open to the future of the revolution.[63] But the realization of just how weighty the past was in this highly traditionbound society is salutary to the effort of understanding how limited the range of alternatives actually was.[64]

As one considers the ways in which earlier social experiences or political decisions shaped those of a later day, it is important to keep in mind that new eras reveal new combinations of experiences and problems—"conjunctures" in the terminology of structuralism.[65] It is no surprise to find that such newly recognized, and even newly formed, conjunctures require unprecedented responses from those individuals who are thought to have the power, authority, or legal responsibility to react to them. For example, in the USSR the combination of an unprecedented demand for administrative-technical cadres at the end of the 1920s with the class principles of the revolution produced the mass-recruitment techniques of the era as a substitute for the elite and sub-elite recruitment patterns of the nineteenth century. Armstrong

63. Cohen, *Rethinking the Soviet Experience*; Lewin, *Political Undercurrents*.
64. Theda Skocpol, "A Critical Review of Barrington Moore's *Social Origins of Dictatorship and Democracy*," *Politics and Society* 4 (1973), 19–34.
65. The term is used, if at all explicitly, in varying senses by social scientists and structuralist historians. Cf. Peter Manicas, "Review Essay: States and Social Revolutions," *History and Theory* 20, no. 2 (1981), 212–13, and the same author's more prolonged discussion (which avoids using the terminology) in "The Concept of Social Structure," *Journal of the Theory of Social Behaviour* 10, no. 2 (1980), 70–76. See also Fernand Braudel's elegant and amusing discussion of conjunctures in economic history in *Civilization and Capitalism, 15th–18th Century*, vol. 3, *The Perspective of the World* (New York, 1984), 71–85.

describes the characteristic of "progressive equal attrition" in the Soviet elite selection process, and his description need not be repeated here.[66] It is important, however, to note that "the principle of plebiscitarianism" (Bendix), "progressive equal attrition" (Armstrong), "mid-career socialization" (Hough and Armstrong) are all terms meant to shed light on fresh solutions to the problem of administrative recruitment in general. Procedures that worked reasonably well in the nineteenth century were no longer feasible because of the fresh combination of circumstances. Many factors made these circumstances unique, but among the most important must surely have been the technical pressure for specialist executives and the sociopolitical pressure for executives drawn from the lowest social classes.

To a remarkable degree, administrative offices changed demographically by the end of the 1920s. If the shift downward in educational qualifications was sharp in the central administration, in the provinces it was sharper at every level. One may also note that the decline in educational qualifications in the provinces of the 1920s was discontinuous. That is, there was not simply a proportionate rise in the number of executive personnel with secondary education equal to the decline in higher education; rather, there was a substantial increase in the proportion of those possessing only a primary or informal (*domashnee*) education, running in many provinces to one-third of the total. Comparable prerevolutionary figures were always negligible at these levels: virtually no one with only a primary or "home" education could expect to attain an important provincial administrative post in the nineteenth century.

In a sense, then, the administrative and managerial generation that the American political scientists Armstrong, Hough, and Bialer discuss—the first generation of "mature Stalinism" in Bialer's curious phrase—represented a fruition of some goals of the revolution perhaps even more than did the generation of 1917. To the extent that they actually came from beyond either the bureaucratic or the upper-class sources that had been common to the previous Russian administration, it was the generation of the 1930s, not the 1920s, that embodied the values of the new society. In this sense it was the first successful product of what, following Lenin, Fitzpatrick calls the Cultural Revolution.[67]

Because one important factor in the formation of any organizational generation is the generation that precedes it, one is inclined to speculate

66. Armstrong, *European Administrative Elite*, 19–23, 244–51.
67. For a discussion of "mature Stalinism," see Bialer, *Stalin's Successors*, 9–21. The "Cultural Revolution" is treated in Sheila Fitzpatrick, *The Russian Revolution, 1917–1932* (New York, 1984), 129–34. See also Fitzpatrick, ed., *Cultural Revolution in Russia, 1928–1931*. (Bloomington, Ind., 1978).

on the impact that the generation of the 1920s may have had on the generation of the 1930s. We have already noted the increasing age difference between elite and lower-level administrators in the 1920s, the class and Party differences, and the differences in education. In fact, of course, these older men were roadblocks until the administrative expansion of the late 1920s and early 1930s, whether they were perceived as such or not. Moreover, the very means by which members of the generation of the 1920s obtained their positions—unorthodox, unexpected, even violent—exhibited at least one path that those of the younger generation could use for their own career advancement. And if advancement for the men—and, for the first time in Russia, women—of the 1920s was justified in the name of revolution, how much more would the revolution justify the advancement of workers and peasants in the 1930s?

Organizational relations may be classified in several ways. They consist not only in relations among colleagues of the same level and in relations between different levels, but also in relations across time. Not only does one generation remember another, its predecessor, but older generations have the power to shape younger ones.Whether they are able to exercise this power depends on the degree to which regular gradual advancement of juniors into senior positions occurs. If such advancement does not occur or if it occurs suddenly or unexpectedly, the intergenerational impact would appear to be impaired and the organization obliged to find other means for training recruits in proper organizational behavior. One may assume that both the generation of the 1920s and that of the 1930s found themselves in the position of having to concentrate organizational resources on training and development of staff to an extraordinary degree and that the cost of this effort was passed on in the form of reduced utility of personnel to the organization. It seems reasonable to believe that midcareer education and indoctrination programs were designed to mitigate the impact on the organization of just such disadvantages.[68]

If I interpret Armstrong and Hough correctly, concerning the installation of the generation of the 1930s and its rapid rise to elite status at the end of the decade, relative stability returned to the ranks of civil administration in the Soviet Union. As both authors note, World War II occasioned some additional, unexpected turnover but not on the order of either of the previous decades. The experiments of the Khrushchev era also produced career instability and jarred intergenerational rela-

---

68. Armstrong, *European Administrative Elite*, 249ff; Hough, *Soviet Leadership*, 48–51.

tions, but again, on a much-reduced scale as compared with the 1920s and 1930s.[69]

With the rise of the new generation of the 1930s, stability among personnel began to return, although at the time this must have been difficult in the extreme, to perceive. The 1930s will always be remembered in Soviet history as a decade of enormous upheaval and even personal danger for many in official employment. And yet the evidence is clear that senior Soviet officialdom, until very recently, was staffed at its core by a cohort that took its rise in the 1930s.[70]

It is important to recognize that it is the graduates of the educational programs of the later 1920s and early 1930s who became the top executives of the later 1930s. By the time the younger generation had assumed its new, commanding role as the "Men of '38" we had witnessed the inauguration of an era of stability that was unprecedented in length; many of these men were still at work until the end of the Brezhnev era.[71] Even Bialer is constrained to remark on the stability of personnel: "The survival ratio of the beneficiaries of the Great Purge during the remainder of the Stalinist era [that is, "mature Stalinism"] was *surprisingly* high."[72] In fact the Men of '38 were not merely the "beneficiaries of the Great Purge." They were the beneficiaries of programs that required enormous administrative expansion; they were the beneficiaries of policies that favored the lower classes for admission to educational institutions, to the Party, and to desirable employment; they were the beneficiaries of the crash education programs of the late 1920s. They were, in short, the heirs, the prime beneficiaries, of the Revolution of 1917. An extraordinary generation, they confounded expectations in their youth just as they did even in very old age.

One may also wonder whether, for the near future, organizational turnover will be "normal" in the USSR. The Men of '38, the generation of the 1930s, were present in significant numbers until very recently. The most successful among them managed to sustain their positions of organizational power and influence for exceptionally long periods of time.[73] One effect of this achievement was a delayed turnover of senior positions after World War II. A certain proportion of the post–World

---

69. George W. Breslauer, *Khrushchev and Brezhnev as Leaders: Building Authority in Soviet Politics* (London, 1982); Sidney Ploss, *Conflict and Decision-Making in Soviet Russia: A Case Study of Agricultural Policy, 1953–1963* (Princeton, N. J., 1965).
70. Armstrong, *European Administrative Elite*, 246–51; Hough, *Soviet Leadership*, chaps. 1, 3; Hodnett, *Leadership*, chap. 4, esp. Table 4.13.
71. Armstrong, *Soviet Bureaucratic Elite*, 26; Hough, *Soviet Prefects*, 38–47; Bialer, *Stalin's Successors*, 63–126.
72. Bialer, *Stalin's Successors*, 89 (my Italics.)
73. Hough, *Soviet Leadership*, chaps. 1, 3; B. Ts. Urlanis, *Istoriia odnogo pokoleniia* (Moscow, 1968).

War II generation, it appears, constituted a pre-elite whose career expectations were frustrated. To put things another way, it is evident that many members of the post–World War II generation were ready for retirement at a time when a lingering portion of the preceding generation—including, until recently, much of the top political leadership—was still in power.

Such a state of affairs suggests a pattern of turnover which, if delayed in the 1960s and 1970s, will be rapid and enduring as the later 1980s give way to the 1990s. The Men of '38 will finally depart the scene, to be followed in comparatively short order by the postwar generation of the 1950s. Thus, at least at senior levels, "normal" generational change in the USSR may in fact not occur for a while longer.

The Emerging Soviet Technostructure: Some Implications

It remains to ask whether there was an enduring legacy from the structural transformation of Soviet administration in the revolutionary era. This is an important and powerful question, focusing attention on the fact that consciously or otherwise, the Soviet leadership exercised certain options—and avoided others—in 1917 and throughout the 1920s which shaped the future of Soviet society. What can one say about the long-term importance of these choices?

First, the creation of specialist administrations confirmed the central role of the state in postrevolutionary society. Although the Russian political leadership had assumed an important role in everything from education to commerce before 1917, the discontinuation of this approach to governance was at least an option in 1917. It is even fair to say that the Provisional Government took a few tentative steps in that direction.

The location of professional and technical management within the precincts of state bureaucracy effectively meant that the revolutionary leadership exercised the statist—or, in the terminology of the day, the statization (*ogosudarstvlenie*)—option.[74] Not only would war, taxes, and police power continue to be the privileged concerns of the state, but so would health, welfare, education, commerce, and economic development. "In this way," as Lewin puts it, "bolshevism acquired a social basis it did not want and did not immediately recognize: the bureaucracy."[75]

Even if—under the New Economic Policy of the 1920s—the political

74. Moshe Lewin, *The Making of the Soviet System* (New York, 1985), 260ff.
75. Ibid., 261.

leadership obliged state administration, the socialized sector of the economy, to share economic control with the private sector in trade, manufacture, and agriculture, the interests of the state always came first. Dzerzhinskii, who as chairman of the Supreme Economic Council until 1926 favored NEP and the role of the market, saw the market as support for state industry—not the other way around.[76] Leadership judged the very viability of NEP in terms of whether it contributed to or impinged upon the freedom of action of the state. And it was on those grounds that the market was finally laid to rest. Beginning at least in 1926, the professional economic planners—economists in the State Planning Commission, for example—and their political supporters gained ground steadily in their fight to constrict the market in favor of the socialized sector.[77]

One must also keep in mind the broader context of state–private sector relations. Throughout the NEP period the role of the private sector, where trade relations were driven by the market, tended to decline as the economy expanded. In 1925–26, out of total trade amounting to Rs 23 billion, the public sector accounted for about 70 percent; in 1928–29, the state's role had already increased to 94 percent of trade, amounting to some Rs 43 billion.[78] Since the era of NEP the interests of the state and the Party have never again been seriously challenged by the public, the market, or, for that matter, any group outside officialdom. Whether, at the end of the 1980s, this period of unquestioned state monopoly is about to end remains to be seen.

Second, the network of actions after 1917 that moved Soviet society toward a technocratic system also implied the permanent bureaucratization of the "free" professions and of the technical intelligentsia. The notion (stated by Medvedev and quoted in the first chapter of this work) that technocracy might in some way be an alternative to bureaucracy is illusory. If bureaucracy is work hierarchically organized and substantively compartmentalized, then it is an essential ingredient of technocracy. Moreover, the quaint Saint-Simonian idea that "true" technocrats will somehow be able to abandon their personal ambitions, their personal interpretation of the appropriate uses to which their expertise should be put, flies in the face of all that we know from our own professional experience and from our personal experiences with bureaucracy. At the very least, we must agree with Bendix that variable, individual circumstances govern the degree to which officials are detached from or interested in the social outcome of their work: "But

76. E. H. Carr and R. W. Davies, *Foundations of a Planned Economy* (New York, 1969). 1 (pt. 2): 625.
77. Ibid., 625–34.
78. Ibid., 635.

bargaining and influence continue to affect the conditions of administrative work and, to this extent, the prevailing bureaucratic pattern is subject to gradual alteration. Whether these alterations are mutually countervailing and hence preserve the identity of bureaucracy, whether they cumulate in one or another direction and give rise to 'neo-patrimonial' or 'neo-feudal' patterns, or whether entirely new types of administration emerge—all this is subject to empirical investigation."[79]

The early Soviet commitment to a specialized administration managed by specialists implied that thereafter the history of the professional and technical intelligentsia thus bureaucratized would always be written in the context both of politics and of bureaucracy. Similarly, this commitment meant that career success within the professions would always be linked to the ability to handle bureaucratic politics—in much the same way that academic success in Western countries implies the ability to handle academic politics effectively.

Third, the movement toward technocracy, beginning in 1917, implied that politics at all levels in the USSR would never again be free of the influence of the mainstream or "official" intelligentsia of Soviet society. Although this characteristic would be slow in asserting its influence on Soviet politics, it would do so irresistibly. To be developed, the economy must be planned—rationally, systematically, comprehensively, and above all professionally. To be part of the development process, agriculture must be integrated logically, comprehensively, and reliably with the rest of the economy. Independent farmers and villages whose reactions to the market were unpredictable were intolerable under such a scheme, just as their local, traditional socialeconomic organizations, the communes, were unacceptable. Never mind the human cost, and never mind that so few competent professionals were available that early efforts at scientific or even orderly management were a caricature of a reliable system. The professional-technical objectives were clear.

Beginning in the 1920s, political decisions with implications for economics, defense, social organization—indeed, political decisions with any broad substance—would have to be taken with due regard (as time passed, with deference) for the scientific-technical cadres and for system. Certainly, episodes would continue to occur where a Stalin or a Khrushchev would act against or in ignorance of the powers of the cadres and the system. Stalin, indeed, would attack the cadres during the 1930s.[80] But he would do so at great and enduring cost.

79. Bendix, *Nation-Building*, 139.
80. J. Arch Getty, *Origins of the Great Purges* (New York, 1985), chap. 6; Bialer, *Stalin's Successors*, pt. 1; Robert Conquest, *The Great Terror: Stalin's Purge of the Thirties*, rev. ed. (New York, 1973), chaps. 8, 9, App. A.

The emerging Soviet technostructure in the early 1920s, moreover, meant that the Soviet official intelligentsia became a power within the state rather than merely a putative tool of the state. Less than half a century after the ascendance of Stalin, even the top political leaders of the state and Party apparatuses were products of the technostructure.

Fourth and finally, the administrative staffing programs of the 1920s continued the tsarist practice of requiring social class qualifications as a prerequisite to administrative appointment. An important, immediate consequence of this policy was that Soviet Russia became the first European country systematically to admit the children of the lowest classes into positions of administrative authority. In the longer term the results of this policy are likely to have been less noticeable because of universal education.

Thus I argue that the events that put the Soviet technostructure on its road to development and administrative authority carried with them fateful implications for the future of Soviet society. Does this mean there were no significant alternatives in the 1920s to history as it transpired? No, it does not.

The shift to a technocratic system was itself an alternative to the pre-1917 generalist administration. One might even argue that continuation of the tsarist statist orientation was a comparable choice by leadership. But like all such choices, these were taken with little apparent appreciation either of the power of constraints or of their long-run implications. Here is the nub of the problem of trying to interpret Soviet history in the light of the *political* alternatives of the 1920s. By focusing so intently on political debates—the conflict between Bukharin and Trotsky, for example—one implies that the "victors" had the capacity to control events much more than they were able historically to do. More important one implies that state-building, the principal task of the revolution, was mainly a political problem.

Stephen Cohen, for example, takes pains to describe the intellectual substance of the Bolshevism of the 1920s as confined to a Left-Right spectrum defined by Trotsky and Bukharin. In a 1985 book he argues that at the end of the 1920s, Stalin took a policy direction antithetical to *both* positions; that is, Stalin's policies were alien to the Bolshevik "main stream" of ideas in the 1920s.[81] One is thus entitled to conclude that it didn't matter who won the debates of the 1920s. Politics took its own course. The anomaly that this conclusion forces is not the only reason Cohen's attempt to attribute the new policy directions of the late 1920s to the political role of Stalin is not convincing. In the end, despite his apparent belief that the USSR had dissolved in social chaos

---

81. Cohen, *Rethinking the Soviet Experience*, chap. 2.

Bureaucratic Structure and Revolution

(he uses Moshe Lewin's term "quicksand society" with heavy emphasis on the disorganized character of the society), even Cohen is obliged to explain that Stalin was supported in his efforts by "enthusiastic agents below": "Zealous officials, intellectuals, workers, and perhaps even some peasants came forward to fight and win on the cultural, industrial, rural, and purge 'fronts,' as they were called. Millions of people were victimized, but millions also benefited from Stalinism and thus identified with it."[82]

As the reader will have gathered, I do not accept a merely political explanation for the advent of Stalinism. The historical evidence does not permit one to conclude that the career of one man, Stalin, was the principal explanatory factor for an enormous social, economic, and political movement. To invoke zealous peasants and workers at the last minute as a kind of tragically misguided cheering section is hardly enough. Even Lewin's view—that the new *apparat* and the network of bosses (*nachal'stvo*) were a natural source of support—though accurate, is too simplistic, both because it excludes broader segments of the society that benefited from the revolution and because it proposes that the *nachal'stvo* was eager to create "its own version of 'autocracy, nationality, and orthodoxy.' "[83] It is unclear, in my opinion, that a new version of "autocracy, nationality, and orthodoxy" was the conscious or unconscious objective of these cadres, just as it is unclear that the political state of Soviet society in the 1930s, the 1960s, or the 1980s will be the final, culminating state of postrevolutionary Russia. The administrative and authority structures of the USSR today, barely two generations after the Revolution of 1917, are still very much in evolution.

82. Ibid., 64, 68.
83. Lewin, *Making of the Soviet System*, 268.

Bibliography

Employment Lists and Directories

All Union Communist Party, Statistical Division. *Kommunisty v sostave apparata gosuchrezhdenii i obshchestvennykh organizatsii. Itogi Vsesoiuznoi Partiinoi perepisi, 1927 goda*. Moscow: Gosudarstvennoe Izdatel'stvo, 1927.
Central Committee, All-Union Communist Party (bolsheviks), Statistical Division. *Sotsial'nyi i natsional'nyi sostav VKP (b): Itogi vsesoiuznoi partiinoi perepisi, 1927 goda*. Moscow: Gosudarstvennoe Izdatel'stvo, 1928.
Central Executive Committee, USSR. *Sostav organov vlasti v Soiuze SSR*. Moscow: OGIZ-IZOGIZ, 1930.
Central Statistical Administration USSR. *Itogi desiatiletiia Sovetskoi vlasti v tsifrakh, 1917–1927 gg*. Moscow: TsSU, 1927.
Ministry of Agriculture. *Spisok lichnogo sostava tsentral'nogo uchrezhdeniia Ministerstva Zemledeliia*. St. Petersburg: Ministerstvo Zemledeliia, 1916.
Ministry of Finances. *Spisok lichnogo sostava Ministerstva Finansov na 1913 g*. St. Petersburg: Ministerstvo Finansov, 1913.
Ministry of Internal Affairs. *Spisok lits sluzhashchikh po vedomstvu Ministerstva Vnutrennikh Del*. St. Petersburg: Tipografiia Ministerstva Vnutrennikh Del, 1907, 1912.
———. *Spisok vysshikh chinov Ministerstva Vnutrennikh Del*. 2 parts (central and provincial lists). St. Petersburg: Tipografiia Ministerstva Vnutrennikh Del, 1896–1916.
———. Medical Department. *Rossiiskii meditsinskii spisok. Spisok Russkikh vrachei na 1912 g*. St. Petersburg: Medical Department, 1912.
Ministry of Trade. *Spisok lichnogo sostava tsentral'nykh uchrezhdenii Ministerstva Torgovli i Promyshlennosti*. St. Petersburg: Ministerstvo Torgovli, 1912.
Ministry of Ways of Communication. *Spisok lichnogo sostava Ministerstva Putei Soobshcheniia*. St. Petersburg: Izdanie Kantseliarii Ministerstva, 1898–1916.

People's Commissariat for Posts and Telegraphs. *Perepis' rabotnikov sviazi, 27 ianvaria, 1927 goda.* Moscow: Izdatel'stvo NKPT, 1929.
People's Commissariat for the Preservation of Health. *Spisok meditsinskikh vrachei SSSR (na 1 ianvaria 1924 g.)* Moscow: NARKOMZDRAV, 1925.
Ruling Senate. *Adres-Kalendar': Obshchaia rospis' nachal'stvuiushchikh i prochikh dolzhnostnykh lits po vsem upravleniia po Rossiiskoi Imperii.* St. Petersburg: Senatskaia Tipografiia, 1904–16.
Supreme Council of National Economy. *Struktura i sostav organov VSNKh v tsentre i na mestakh v 1921 g.* Moscow: VSNKh, 1922.
Vsia Moskva: Adresnaia i spravochnaia kniga. Moscow: Gosudarstvennoe Izdatel'stvo, 1923, 1927, 1936.

Books

Abrahamson, Mark, ed. *The Professional in the Organiztion.* Chicago: Rand McNally, 1967.
Alekseeva, G. *Oktiabr'skaia revoliutsiia i istoricheskaia nauka, 1917–1923.* Moscow: Nauka, 1968.
Amburger, Erik. *Geschichte der Behördenorganisation Russlands vom Peter dem Grossen bis 1917.* Leiden: Brill, 1966.
Anderson, Barbara. *Internal Migration during Modernization in Late Nineteenth-Century Russia.* Princeton, N.J.: Princeton University Press, 1980.
Anderson, Perry. *Lineages of the Absolutist State.* London: Verso, 1974.
Andreev, A. M. *Mestnye sovety i organy burzhuaznoi vlasti (1917 g.).* Moscow: Nauka, 1983.
Anweiler, Oskar. *The Soviets: The Russian Workers, Peasants, and Soldiers Councils, 1905–1921.* New York: Pantheon, 1974.
Armstrong, John A. *The European Administrative Elite.* Princeton, N.J.: Princeton University Press, 1973.
———. *The Soviet Bureaucratic Elite: A Case Study of the Ukrainian Apparatus.* New York: Praeger, 1959.
Ashton-Tate, Inc. *Programming with dBase III Plus.* Torrence, Calif., 1987.
Atkinson, Dorothy. *The End of the Russian Land Commune, 1905–1930.* Stanford, Calif.: Stanford University Press, 1983.
Azrael, Jeremy. *Managerial Power and Soviet Politics.* Cambridge, Mass.: Harvard University Press, 1966.
Bailes, Kendall E. *Technology and Society under Lenin and Stalin: Origins of the Soviet Technical Intelligentsia, 1917–1941.* Princeton, N.J.: Princeton University Press, 1978.
Bairoch, Paul, T. Deldycke, H. Gelders, and J.-M. Limbor. *International Historical Statistics.* Brussels: Kluwer, 1973.
Barber, John. *Soviet Historians in Crisis, 1928–1932.* New York: Holmes & Meyer, 1981.
Barker, Ernest. *The Development of Public Services in Western Europe.* New York: Oxford University Press, 1944.
Beck, Carl, and J. Thomas McKechnie. *Political Elites: A Select, Computerized Bibliography.* Cambridge, Mass.: MIT Press, 1968.
Becker, Seymour. *Nobility and Privilege in Late Imperial Russia.* DeKalb: Northern Illinois University Press, 1985.
Bendix, Reinhard. *Kings or Peoples: Power and the Mandate to Rule.* Berkeley: University of California Press, 1978.

Bibliography

———.*Nation-Building and Citizenship: Studies of Our Changing Social Order*. Rev. ed. Berkeley: University of California Press, 1977.

———.*Work and Authority in Industry: Ideologies and Management in the Course of Industrialization*. New York: Harper & Row, 1963.

Bettelheim, Charles H. *Class Struggles in the USSR*. Vol. 1, *First Period, 1917–1923*; Vol. 2, *Second Period, 1923–1930*. New York: Monthly Review, 1976, 1978.

Bialer, Seweryn. *Stalin's Successors: Leadership, Stability, and Change in the Soviet Union*. New York: Cambridge University Press, 1980.

Bineman, Ia., and S. Kheinman. *Kadry gosudarstvennogo i kooperativnogo apparata SSSR*. Moscow: Planovo-khoziaistvennoe Izdatel'stvo, 1930.

Blum, Jerome. *The End of the Old Order in Rural Europe*. Princeton, N.J.: Princeton University Press, 1978

———. *Lord and Peasant in Russia from the Ninth to the Nineteenth Century*. New York: Atheneum, 1964.

Bol'shaia sovetskaia entsiklopediia. 1st ed. Moscow, 1926–1947.

Braudel, Fernand. *Capitalism and Material Life, 1400–1800*. New York: Harper & Row, 1967.

———*Civilization and Capitalism, 15th–18th Century*. 3 vols. Trans. Sian Reynolds. New York: Harper & Row, 1979.

Breslauer, George W. *Krushchev and Brezhnev as Leaders: Building Authority in Soviet Politics*. London: Allen & Unwin, 1982.

Brode, John. *The Process of Modernization: An Annotated Bibliography on the Sociocultural Aspect of Development*. Cambridge, Mass.: Harvard University Press, 1969.

Brzezinski, Zbigniew, and Samuel P. Huntington. *Political Power: USA/USSR*. New York: Viking Press, 1963.

Carr, E. H. *The October Revolution: Before and After*. New York: Vintage, 1971.

———. *The Russian Revolution from Lenin to Stalin, 1917–1929*. London: Macmillan, 1979.

———. *Socialism in One Country, 1924–1926*. Vol. 3 of *History of Soviet Russia*. London: Macmillan, 1964.

Carr, E. H., and Robert W. Davies. *Foundations of a Planned Economy, 1926–1929*. 2 vols. New York: Macmillan, 1969–71.

Central Statistical Administration, *Statisticheskii ezhegodnik Rossii*. St. Petersburg: Izdatel'stvo Tsentral'nogo Statisticheskogo Upravleniia, 1893–1914.

Coale, Ainsley J., Barbara A. Anderson, and Erna Harm. *Human Fertility in Russia since the Nineteenth Century*. Princeton, N.J.: Princeton University Press, 1979.

Cohen, Stephen. *Bukharin and the Bolshevik Revolution: A Political Biography, 1888–1938*. New York: Knopf, 1973.

———. *Rethinking the Soviet Experience*. New York: Oxford University Press, 1985.

Conquest, Robert. *The Great Terror: Stalin's Purge of the Thirties*. Rev. ed. New York: Collier, 1973.

———. *V. I. Lenin*. New York: Viking, 1972.

Crisp, Olga. *Studies in the Russian Economy before 1914*. London: Macmillan, 1976.

Crowley, E. L., ed. *Party and Government: Officials of the Soviet Union, 1917–1967*. Metuchen, N.J.: Scarecrow Press, 1969.

Crozier, Michel. *The Bureaucratic Phenomenon*. Chicago: Rand-McNally, 1964.

Daniels, Robert V. *The Conscience of the Revolution*. Cambridge, Mass.: Harvard University Press, 1960.

Dashkevich, Leonid. *Nashe Ministerstvo Vnutrennikh Del*. Berlin, 1885.

Davies, R. W. *The Socialist Offensive: The Collectivisation of Soviet Agriculture*,

1929–1930. Vol. 2 of *The Industrialisation of Soviet Russia*. Cambridge, Mass.: Harvard University Press, 1980.
Day, Richard B. *Leon Trotsky and the Politics of Economic Isolation*. Cambridge: Cambridge University Press, 1973.
Deutscher, Isaac. *The Prophet Armed: Trotsky, 1879–1921*. New York: Vintage, 1965.
———. *The Prophet Outcast: Trotsky, 1928–1940*. New York: Vintage, 1965.
———. *The Prophet Unarmed: Trotsky: 1921–1929*. New York: Vintage, 1965.
Drobizhev, V. Z. *Glavnyi shtab sotsialisticheskoi promyshlennosti*. Moscow: Nauka, 1966.
Dubrovskii, S. M. *Krest'ianstvo v 1917 gody*. Moscow-Leningrad: Gosudarstvennoe Izdatel'stvo, 1927
Eisenstadt, S. N. *The Political System of Empires: The Rise and Fall of Historical Bureaucratic Societies*. New York: Free Press, 1963.
———. *Revolution and the Transformation of Societies: A Comparative Study of Civilizations*. New York: Free Press, 1978.
Emmons, Terrence. *The Russian Landed Gentry and the Peasant Emancipation of 1861*. Cambridge: Cambridge University Press, 1968.
Eroshkin, N. P. *Istoriia gosudarstvennykh uchrezhdenii dorevoliutsionnoi Rossii*. 3d rev. ed. Moscow: Vysshaia Shkola, 1983.
Fainsod, Merle. *Smolensk under Soviet Rule*. New York: Vintage, 1958.
Fediukin, S. A. *Privlechenie burzhuaznoi tekhnicheskoi intelligentsii k sotsialisticheskomu stroitel'stvu v SSSR*. Moscow: Nauka, 1960.
———. *Sovetskaia vlast' i burzhuaznye spetsialisty*. Moscow: Nauka, 1972.
Feit, Edward. *The Armed Bureaucrats*. Boston: Houghton Mifflin, 1973.
Ferro, Marc. *Des soviets au communisme bureaucratique: Les mecanismes d'une subversion*. Paris: Gallimard/Juillard, 1980.
———. *October, 1917: A Social History of the Russian Revolution*. London: Routlege & Kegan Paul, 1980.
Fitzpatrick, Sheila. *The Commissariat of Enlightenment: Soviet Organization of Education and the Arts under Lunacharsky*. Cambridge: Cambridge University Press, 1970.
———. *Education and Social Mobility in the Soviet Union, 1921–1934*. Cambridge: Cambridge University Press, 1979.
———. *The Russian Revolution, 1917–1932*. New York: Oxford University Press, 1984.
———. ed. *Cultural Revolution in Russia, 1928–1931*. Bloomington: Indiana University Press, 1984.
Forrester, Jay W. *Principles of Systems*. 2d prelim. ed. 2 vols. Cambridge, Mass.: Wright-Allen, 1968.
———. *Urban Dynamics*. Cambridge, Mass.: MIT Press, 1969.
Friedrich, C. J., and Zbigniew K. Brzezinski. *Totalitarian Dictatorship and Autocracy*. Cambridge, Mass.: Harvard University Press, 1956.
Gavrilov, L. M. *Soldatskie komitety v oktiabr'skoi revoliutsii*. Moscow: Nauka, 1983.
Gerasimenko, G. A. *Nizovye krest'ianskie organizatsii v 1917—pervoi polovine 1918 godov: Na materialakh nizhnego Povolzh'ia*. Saratov: Saratov University Press, 1974.
Gershenkron, Alexander. *Continuity in History and Other Essays*. Cambridge, Mass.: Harvard University Press, 1968.
Getty, J. Arch. *Origins of the Great Purges: The Soviet Communist Party Reconsidered, 1933–1938*. New York: Cambridge University Press, 1985.
Gimpel'son, E. G. *Rabochii klass v upravlenii sovetskim gosudarstvom: Noiabr' 1917–1920 gg*. Moscow: Nauka, 1982.

———. *Velikii Oktiabr' i stanovlenie sovetskoi sistemy upravleniia narodnym khoziaistvom (Noiabr' 1917–1920 gg.)*. Moscow: Nauka, 1977.
Gorodetskii, E. N. *Rozhdenie sovetskogo gosudarstva, 1917–1918 gg.* Moscow: Nauka, 1965.
Gosudarstvennyi apparat SSSR, 1924–1928. Moscow, 1929.
Gouldner, Alvin W. *Patterns of Industrial Bureaucracy*. Glencoe, Ill.: Free Press, 1954.
———. ed. *Studies in Leadership: Leadership and Democratic Action*. New York: Wiley, 1950.
Granick, David. *The Red Executive: A Study of the Organization Man in Russian Industry*. New York: Doubleday, 1960.
Haimson, Leopold, ed. *The Politics of Rural Russia, 1905–1914*. Bloomington: Indiana University Press, 1979.
Hamburg, G. M. *Politics of the Russian Nobility, 1881–1905*. New Brunswick, N.J.: Rutgers University Press, 1984.
Haupt, Georg, and Jean-Jacques Marie. *Makers of the Russian Revolution: Biographies of Bolshevik Leaders*. London: Allen & Unwin, 1974.
Hodnett, Grey. *Leadership in the Soviet National Republics: A Quantitative Study of Recruitment Policy*. Oakville, Ont.: Mosaic Press, 1978.
Hoffmann, Eric P., and Robbin F. Laird. *The Politics of Economic Modernization in the Soviet Union*. Ithaca: Cornell University Press, 1982.
Hough, Jerry F. *Soviet Leadership in Transition*. Washington, D.C.: Brookings Institution, 1980.
———. *The Soviet Prefects: The Local Party Organs in Industrial Decision-Making*. Cambridge, Mass.: Harvard University Press, 1969.
———. *The Soviet Union and Social Science Theory*. Cambridge, Mass.: Harvard University Press, 1977.
Huntington, Samuel P. *Political Order in Changing Societies*. New Haven, Conn.: Yale University Press, 1968.
Inkeles, Alex. *Social Change in Soviet Russia*. Cambridge, Mass.: Harvard University Press, 1968.
Institute of Marxism-Leninism under the Central Committee of the Communist Party of the Soviet Union. *KPSS v rezoliutsiiakh i resheniiakh s"ezdov, konferentsii i plenumov TsK, 1898–1969*. 7th ed. 4 vols. Moscow: Gosudarstvennoe Izdatel'stvo Politicheskoi Literatury, 1954–60.
Iroshnikov, M. P. *Sozdanie Sovetskogo tsentral'nogo gosudarstvennogo apparata: Sovet Narodnykh Komissarov i Narodnye Komissariaty, Oktiabr' 1917 g.–Ianvar' 1918 g.* 2d rev. Leningrad: Nauka, 1967.
Ivanova, L. V. *Formirovanie sovetskoi nauchnoi intelligentsii, 1917–1927*. Moscow: Nauka, 1980.
Ivanovskii, V. V. *Uchebnik administrativnogo prava (Politseiskoe pravo: Pravo vnutrennago upravleniia)*. 3d ed. Kazan: Imperial University of Kazan, 1908.
Joravsky, David. *The Lysenko Affair*. Cambridge, Mass.: Harvard University Press, 1970.
Judge, Edward H. *Plehve: Repression and Reform in Imperial Russia, 1902–1904*. Syracuse, N.Y.: Syracuse University Press, 1983.
Kabuzan, V. M. *Narodonaselenie Rossii v XVIII-pervoi polovine XIX v. (Po materialam revizii)*. Moscow: Nauka, 1963.
Karamzin. See Pipes, Richard, ed.
Kautsky, John H. *Political Change in Underdeveloped Countries: Nationalism and Communism*. New York: Wiley, 1962.

Keep, John L. H. *The Russian Revolution: A Study in Mass Mobilization.* New York: Norton, 1976.
——. ed. *The Debate on Soviet Power: Minutes of the All-Russian Central Executive Committee of Soviets, Second Session.* Oxford: Oxford University Press, 1979.
Kenez, Peter. *The Birth of the Propaganda State: Soviet Methods of Mass Mobilization, 1917–1929.* Cambridge: Cambridge University Press, 1985.
Kitanina, T. M. *Khlebnaia torgovlia Rossii v 1875–1914 gg.* Leningrad: Nauka, 1978.
Koenker, Diane. *Moscow Workers and the 1917 Revolution.* Princeton, N.J.: Princeton University Press, 1981.
Korelin, A. P. *Dvorianstvo v poreformennoi Rossii: Moscow: Nauka, 1860–1904.* Moscow: Nauka, 1979.
Korkunov, N. M. *Russkoe gosudarstvennoe pravo.* St. Petersburg, 1909.
Korzhikhina, T. P. *Istoriia gosudarstvennykh uchrezhdenii SSSR.* Moscow: Vysshaia Shkola, 1986.
Kotel'nikov, K. G., and V. L. Meller. *Krest'ianskoe dvizhenie v 1917 godu.* Ed. M. N. Pokrovskii and Ia. L. Iakovlev. Arkhiv Oktiabr'skoi Revoliutsii 1917 g. v dokumentakh i materialakh. Moscow: Gosudarstvennoe Izdatel'stvo, 1927.
Koval'chenko, I. D., and L.V. Milov. *Vserossiiskii agrarnyi rynok XVIII-nachalo XX veka: Opyt kolichestvennogo analiza.* Moscow: Nauka, 1974.
Koval'chenko, I. D., N. B. Selunskaia, and B. M. Litvakov. *Sotsial'no-ekonomicheskii stroi pomeshchich'ego khoziaistva evropeiskoi Rossii v epokhu kapitalizma.* Moscow: Nauka, 1982.
Kriashcheva, A. I. *Gruppy i klassy v krest'ianstve—k XIII s'ezdu RKP (b).* 1st and 2d eds. Moscow: Tsentral'noe statisticheskoe upravleniie, 1924, 1926.
Kuz'min-Karavaev, V. D. *Zemstvo i Derevnia, 1898–1908. Stat'i, referaty, doklady i rechi.* N.p.: Biblioteka Obshchestvennoi Pol'zy, n.d.
Kuznets, Simon. *Economic Growth and Structure.* New York: Norton, 1965.
Kuznetsov, A. A. *Ordena i medali Rossii.* Moscow: Izdatel'stvo Moskovskogo Universiteta, 1985.
Lampert, Nicholas. *The Technical Intelligentsia and the Soviet State.* New York: Holmes & Meier, 1979.
League of Nations, Health Section, Epidemiological Intelligence. *Epidemics in Russia since 1914: Report to the Health Committee of the League of Nations by Professor L. Tarassevitch (Moscow). Pt. 1, Typhus—Relapsing Fever—Smallpox.* Geneva: League of Nations, 1922.
Lenin, V. I. *Polnoe sobranie sochinenii.* 5th ed. 55 vols. Moscow: Gosudarstvennoe Izdatel'stvo Politicheskoi Literatury, 1958–65.
Lesnoi, V. M. *Istoriia sovetskogo gosudarstva i prava (1917–1920 gg.).* Moscow: Nauka, 1968.
——. *Sovety v pervyi god proletarskoi diktatury, Oktiabr' 1917 g.—Noiabr' 1918 g.* Moscow: Nauka, 1967.
——. ed. *Sotsialisticheskaia revoliutsiia i gosudarstvennyi apparat.* Moscow: Nauka, 1968.
Levi-Strauss, Claude. *The Raw and the Cooked.* Trans. John and Doreen Weightman. Science and Mythology Series, Vol. 1. New York: Octagon, 1970.
——. *Structural Anthropology.* Trans. Claire Jacobsen and Brooke G. Schoepf. New York: Basic Books/Harper, 1963.
——. *Tristes Tropiques.* Trans. John and Doreen Weightman. New York: Atheneum, 1975.
——. *The View from Afar.* Trans. Joachim Neugroschel and Phoebe Hoss. New York: Basic Books, 1985.

Bibliography 217

Lewin, Moshe. *The Making of the Soviet System: Essays in the Social History of Interwar Russia*. New York: Pantheon, 1985.
——. *Political Undercurrents in Soviet Economic Debates: From Bukharin to the Modern Reformers*. Princeton, N.J.: Princeton University Press, 1974.
——. *Russian Peasants and Soviet Power: A Study of Collectivization*. New York: Norton, 1968.
Liashchenko, P. I. *History of the National Economy of Russia*. New York: Octagon, 1970.
Lincoln, W. Bruce. *Nikolai Miliutin: An Enlightened Russian Bureaucrat*. Newtonville, Mass.: Oriental Research Partners, 1977.
Lorimer, Frank. *Population of the Soviet Union: History and Prospects*. Geneva: League of Nations, 1946.
Lotus Development Corporation. *Lotus 123: Reference*. Cambridge, Mass.: Lotus, 1987.
McGrew, Roderick E. *Russia and the Cholera, 1823–1832*. Madison: University of Wisconsin Press, 1965.
Macksey, Richard, and Eugenio Donato, eds. *The Structuralist Controversy: The Languages of Criticism and the Sciences of Man*. Baltimore, Md.: Johns Hopkins University Press, 1970.
McNeal, Robert H. *Guide to the Decisions of the Communist Party of the Soviet Union, 1917–1967*. Toronto: University of Toronto Press.
McNeill, William H. *The Pursuit of Power: Technology, Armed Force, and Society since A.D. 1000*. Chicago: University of Chicago Press, 1982.
Maine, Henry. *Ancient Law*. New York: Everyman's Library, 1931.
Male, D. J. *Russian Peasant Organisation before Collectivisation: A Study of Commune and Gathering, 1925–1930*. Cambridge: Cambridge University Press, 1971.
Maliavskii, A. D. *Krest'ianskoe dvizhenie v Rossii v 1917 g. Mart-Oktiabr'*. Moscow: Nauka, 1981.
Manning, Roberta T. *The Crisis of the Old Order in Russia: Gentry and Government*. Princeton, N.J.: Princeton University Press, 1982.
Mayer, Arno J. *Dynamics of Counterrevolution in Europe, 1870–1956: An Analytic Framework*. New York: Harper & Row, 1971.
——. *The Persistence of the Old Regime, Europe to the Great War*. New York: Pantheon, 1981.
Medvedev, Roy. *All Stalin's Men: Six Who Carried Out the Bloody Policies*. New York: Anchor, 1985.
——. *On Socialist Democracy*. New York: Norton, 1975.
Melidov, A. A. *Istoriia gosudarstvennykh uchrezhdenii SSSR, 1917–1936 gg.: Uchebnoe posobie*. Moscow: Vysshaia Shkola, 1962.
Miller, Margaret. *The Economic Development of Russia, 1905–1914*. 2d ed. New York: Augustus M. Kelley Reprints, 1967.
Ministry of Internal Affairs. Medical Department. *Otchet Meditsinskogo Departamenta Ministerstva Vnutrennikh Del za 1891 god*. St. Petersburg: Tipografiia Ministerstva Vnutrennikh Del, 1891.
——. Medical Department. *Otchet o sostoianii narodnogo zdraviia i organizatsii vrachebnoi pomoshchi v Rossii za 1907 god*. St. Petersburgh: Tipografiia Ministerstva Vnutrennikh Del, 1909.
——. *Smeta dokhodov i raskhodov i spetsial'nykh sredstv Ministerstva Vnutrennikh Del na 1905 g*. St. Petersburg: Izdatel'stvo Ministerstva Vnutrennikh Del, 1904.
Mitchell, B. R. *European Historical Statistics, 1750–1970*. London: Macmillan, 1975.

Moore, Barrington, Jr. *Social Origins of Dictatorship and Democracy: Lord and Peasant in the Making of the Modern World*. Boston: Beacon, 1966.

——. *Terror and Progress—USSR: Some Sources of Change and Stability in the Soviet Dictatorship*. Cambridge, Mass.: Harvard University Press, 1954.

Northwest Analytical, Inc. *NWA Statpak: Multi-Function Statistics Library*. Version 3.1. Portland, Ore.: 1984.

Novik, E. K. *Formirovanie kadrov narodnogo obrazovaniia Belorussii (1917–1941 gg.)*. Minsk: Nauka i Tekhnika, 1981.

Odom, William E. *The Soviet Volunteers: Modernization and Bureaucracy in a Public Mass Organization*. Princeton, N.J.: Princeton University Press, 1973.

Orlovsky, Daniel T. *The Limits of Reform: The Ministry of Internal Affairs in Imperial Russia, 1802–1881*. Cambridge, Mass.: Harvard University Press, 1981.

Owen, Launcelot A. *The Russian Peasant Movement, 1906–1917*. London: S. King, 1937.

Paige, Jeffrey M. *Agrarian Revolution: Social Movements and Export Agriculture in the Underdeveloped World*. New York: Free Press, 1975.

Pascal, Pierre. *En Communisme: Mon Journal de Russie, 1918–1921*. Lausanne: L'Age d'Homme, 1977.

——. *Russie 1927: Mon Journal de Russie*. Lausanne: L'Age d'Homme, 1982.

People's Commissariat for the Preservation of Health. *Izvestiia sovetskoi meditsiny*. Moscow: Izdatel' stvo NARKOMZDRAVa, 1918.

Pethybridge, Roger. *The Social Prelude to Stalinism*. New York: St. Martin's Press, 1974.

Pintner, Walter M. *Russian Economic Policy under Nicholas I*. Ithaca: Cornell University Press, 1967.

Pintner, Walter M., and Don Karl Rowney, eds. *Russian Officialdom: The Bureaucratization of Russian Society from the Seventeenth to the Twentieth Century*. Chapel Hill: University of North Carolina Press, 1980.

Pipes, Richard, ed. *Karamzin's Memoir on Ancient and Modern Russia*. Cambridge, Mass.: Harvard University Press, 1966.

Ploss, Sidney. *Conflict and Decision-Making in Soviet Russia: A Case Study of Agricultural Policy, 1953–1963*. Princeton, N.J.: Princeton University Press, 1965.

Poliakov, Iu. A. *Sovetskaia strana posle okonchaniia grazhdanskoi voiny: territoriia i naselenie*. Moscow: Nauka, 1986.

Pospelov, P. N., V. E. Evgrafov, V. Ia. Zevin, L. F. Il'ichev. *Vladimir Il'ich Lenin: Biografiia*. Moscow: Gosispolit, 1960.

Potulov, V. M. *V. I. Lenin i okhrana zdrov'ia sovetskogo narod*. Leningrad: Meditsina, 1969.

Poulantzas, Nicos. *Classes in Contemporary Capitalism*. 2d ed. London: New Left Books, 1978.

——. *Political Power and Social Classes*. London: New Left Books, 1973.

Proskuriakova, N. A., and A. P. Korelin. *Materialy po istorii agrarnykh otnoshenii v Rossii v kontse XIX—nachale XX vv: Statistika dolgosrochnogo kredita v Rossii*. Moscow: Nauka, 1980.

Rabinowitch, Alexander. *The Bolsheviks Come to Power: The Revolution of 1917 in Petrograd*. New York: Norton, 1976.

Raeff, Marc. *Michael Speransky, Statesman of Imperial Russia, 1772–1839*. The Hague: Nijhoff, 1957.

——. *Origins of the Russian Intelligentsia: The Eighteenth-Century Nobility*. New York: Harcourt, Brace & World, 1966.

Rashin, A. G. *Naselenie Rossii za 100 let (1811–1913): Statisticheskie ocherki*. Moscow: Gosudarstvennoe Statisticheskoe Izdatel'stvo, 1956.

Bibliography

Rein, G. E. *Iz perezhitogo, 1907–1918.* 2 vols. Berlin: Parabola, 1937.
Riabyshkin, T. B., V. M. Simchera, et al., eds. *Statisticheskie metody v upravlenii.* Moscow: Nauka, 1980.
Rieber, Alfred J. *Merchants and Entrepreneurs in Imperial Russia.* Chapel Hill: University of North Carolina Press, 1982.
Rigby, T. Harry. *Communist Party Membership in the USSR, 1917–1967.* Princeton, N.J.: Princeton University Press, 1968.
——. *Lenin's Government: SOVNARKOM, 1917–1922.* Cambridge: Cambridge University Press, 1979.
Robbins, Richard G., Jr. *Famine in Russia, 1891–1892: The Imperial Government Responds to a Crisis.* New York: Columbia University Press, 1975.
Robinson, Geroid T. *Rural Russia under the Old Regime.* New York: Macmillan, 1932.
Rodionova, E. I. *Ocherki istorii professional'nogo dvizheniia meditsinskikh rabotnikov.* Moscow: Izdatel'stvo Meditsinskoi Literatury, 1962.
Rowney, Don Karl, ed. *Soviet Quantitative History.* Los Angeles: SAGE, 1983.
Rowney, Don Karl, and James Q. Graham, eds. *Quantitative History: Selected Readings in the Quantitative Analysis of Historical Data.* Homewood, Ill.: Dorsey, 1969.
Ryan, John. *Marxism and Deconstruction.* Baltimore, Md.: Johns Hopkins University Press, 1980.
Schapiro, Leonard. *The Communist Party of the Soviet Union.* 2d ed. New York: Random House, 1970.
——. *The Origin of the Communist Autocracy: Political Opposition in the Soviet State, First Phase, 1917–1922.* Cambridge, Mass.: Harvard University Press, 1956.
Selznik, Philip. *The Organizational Weapon: A Study of Bolshevik Strategy and Tactics.* Glencoe, Ill.: Free Press, 1960.
——. *TVA and the Grass Roots.* Berkeley: University of California Press, 1949.
Service, Robert. *The Bolshevik Party in Revolution: A Study in Organizational Change.* London: Macmillan, 1979.
Sfez, Lucien. *Critique de la Decision.* Paris: Presses de la Fondation Nationale des Sciences Politiques, 1981.
Shanin, Teodor. *The Awkward Class: Political Sociology of Peasantry in a Developing Society: Russia, 1910–1925.* Oxford: Clarendon Press, 1972.
——. *The Roots of Otherness: Russia's Turn of the Century.* London: Macmillan, 1986.
——. *Russia as a Developing Society.* London: Macmillan, 1985.
Sherstobitov, V. P., I. M. Volkov, R. Kh. Aminova, Iu. V. Arutiunian. *Istoriia krest'ianstva SSSR: Krest'ianstvo v pervoe desiatiletie Sovetskoi vlasti, 1917–1927.* Vol. 1. Moscow: Nauka, 1986.
Silberman, Bernard. *Ministers of Modernization: Elite Mobility in the Meiji Restoration, 1868–1873.* Tucson: University of Arizona Press, 1964.
Simerenko, Alex. *Professionalization of Soviet Society.* Ed. C.A. Kern-Simerenko. New Brunswick, N.J.: Transaction Books, 1982.
Simon, Herbert. *Administrative Behavior.* New York: Wiley, 1948.
Simonova, M. S. *Krizis agrarnoi politiki tsarizma nakanune pervoi rossiiskoi revoliutsii.* Moscow: Nauka, 1987.
Skocpol, Theda. *States and Social Revolutions: A Comparatiave Analysis of France, Russia, and China.* New York: Cambridge University Press, 1979.
Smith, S. A. *Red Petrograd: Revolution in the Factories, 1917–1918.* Cambridge: Cambridge University Press, 1983.
Smitten, E. *Sostav vsesoiuznoi kommunisticheskoi partii (bol'shevikov): Po mater-*

ialam partiinoi perepisi, 1927 goda. Moscow-Leningrad: Gosudarstvennoe Izdatel'stvo, 1927.
Sorokin, Pitirim A. *Leaves from a Russian Diary—and Thirty Years After.* Boston: Beacon Press, 1950.
———. *The Sociology of Revolution.* Philadelphia: Lippincott, 1925.
Speer, Albert. *Infiltration.* New York: Macmillan, 1981.
SPSS, Inc. *SPSS X User's Guide.* New York: McGraw-Hill, 1983.
Starr, S. Frederick. *Decentralization and Self-Government in Russia, 1830–70.* Princeton, N.J.: Princeton University Press, 1972.
Startsev, V. I. *Vnutrennaiaia politika Vremennogo Pravitel'stva.* Leningrad: Nauka, 1980.
Stone, Lawrence, and Jeanne F. Stone. *An Open Elite? England, 1540–1880.* Oxford: Oxford University Press, 1984.
Suleiman, Ezra N. *Elites in French Society: The Politics of Survival.* Princeton, N.J.: Princeton University Press, 1978.
———. *Politics, Power, and Bureaucracy in France.* Princeton, N.J.: Princeton University Press, 1974.
Talmon, J. L. *The Origins of Totalitarian Democracy.* New York: Praeger, 1960.
Theonig, Jean-Claude. *L'ère des technocrates: Le cas des Ponts et Chaussées.* Paris: Editions d'Organisation, 1973.
Therborn, Göran. *Science, Class, and Society: On the Formation of Sociology and Historical Materialism.* London: Verso, 1980.
Trimberger, Ellen Kay. *Revolution from Above: Military Bureaucrats and Development in Japan, Turkey, Egypt, and Peru.* New Brunswick, N.J.: Transaction Books, 1979.
Troitskii, S. M. *Russkii absoliutizm i dvorianstvo v XVIII v.* Moscow: Nauka, 1974.
Trotsky, L. D. *Kak vooruzhalas' revoliutsiia.* Moscow: 1923.
———. *The Revolution Betrayed.* New York: Doubleday, 1937.
———. *Stalin: An Appraisal of the Man and His Influence.* Trans. and ed. Charles Malamuth. New York: Stein & Day, 1967.
Tufte, Edward R. *The Visual Display of Quantitative Information.* Cheshire, Conn.: Graphics Press, 1983.
Tugan-Baranovskii, M. *Russkaia fabrika v proshlom i nastoiashchem.* 7th ed. 2 vols. Moscow: Gosudarstvenoe-Ekonomicheskoe Izdatel'stvo, 1938.
Ulam, Adam B. *Stalin: The Man and His Era.* New York: Viking, 1973.
Ulianovskaia, V. A. *Formirovanie nauchnoi intelligentsii v SSSR, 1917–1937 gg.* Moscow: Nauka, 1966.
Urlanis, B. Ts. *Istoriia odnogo pokoleniia.* Moscow: Mysl', 1968.
Vasiaev, V. I., V. Z. Drobizhev, L. V. Zaks, E. I. Pivovar, V. A. Ustinov, and T. A. Ushakova. *Dannye perepisi sluzhashchikh 1922 g. O Sostave Kadrov NARKOMATov RSFSR.* Moscow: Moscow State University, 1972.
Vestnik vremennogo pravitel'stva. Petrograd, 5 March–27 October 1917.
Vishnevskii, A. G. *Brachnost', rozhdaemost', smertnost' v Rossii i v SSSR: Sbornik statei.* Moscow: Statistika, 1977.
Volin, Lazar. *A Century of Russian Agriculture: From Alexander II to Khrushchev.* Cambridge, Mass.: Harvard University Press, 1965.
Volkov, E. Z. *Dinamika narodonaseleniia SSSR za vosem' desiat let.* Moscow, 1930.
Von Laue, Theodore H. *Sergei Witte and the Industrialization of Russia.* New York: Atheneum, 1969.
———. *Why Lenin? Why Stalin? A Reappraisal of the Russian Revolution, 1900–1930.* Philadelphia: Lippincott, 1964.

Weber, Max. *Law in Economy and Society*. Trans. and ed. Max Rheinstein and E. A. Shils. Cambridge, Mass.: Harvard University Press, 1954.
———. *The Theory of Social and Economic Organization*. New York: Oxford University Press, 1947.
Weissman, Neil B. *Reform in Tsarist Russia: The State Bureaucracy and Local Government, 1900–1914*. New Brunswick, N.J.: Rutgers University Press, 1981.
Wheatcroft, S. W., and Robert W. Davies. *Materials for a Balance of the Soviet National Economy, 1928–1930*. Cambridge: Cambridge University Press, 1985.
Who Was Who in the USSR. Methuen, N.J.: Scarecrow Press, 1972.
Wildman, Allan K. *The End of the Russian Imperial Army: The Old Army and the Soldiers' Revolt*. Princeton, N.J.: Princeton University Press, 1980.
———. *The Making of a Workers' Revolution: Russian Social Democracy, 1891–1903*. Chicago: University of Chicago Press, 1967.
Wolf, Eric R. *Peasant Wars of the Twentieth Century*. New York: Harper & Row, 1969.
Wortman, Richard. *The Development of a Russian Legal Consciousness*. Chicago: University of Chicago Press, 1976.
Yaney, George. *The Systematization of Russian Government*. Urbana: University of Illinois Press, 1973.
Zaionchkovskii, P. A. *Otmena krepostnogo prava v Rossii*. Moscow: Mysl', 1960.
———. *Pravitel'stvennyi apparat samoderzhavnoi Rossii v XIX v*. Moscow: Nauka, 1978.
———. *Spravochniki po istorii dorevoliutsionnoi Rossii: Bibliografiia*. Moscow: Kniga, 1971.

Articles and Dissertations

Abellan, Angel-Manuel. "From Liberal Bureaucracy to Saint-Simonian Bureaucràtic Ideology." *Revisita de estudios politicas* (Spain) 29 (1982), 197–208.
Abrams, Philip. "History, Sociology, Historical Sociology." *Past and Present* 87 (1980), 3–16.
Afanas'ev, V. G. "V. I. Lenin o nauchnom upravlenii obshestvom." *Voprosii filosofii*, no. 1 (1974), 17–29.
Alchon, Guy. "Technocratic Social Science and the Rise of Managed Capitalism, 1910–1933." Ph.D. diss. University of Iowa, 1982.
Andreev, A. M. "Bol'sheviki i mestnye organy burzhuaznoi vlasti v period mirnogo razvitiia Revoliutsii." *Voprosy istorii KPSS* 8 (1979), 71–79.
Armstrong, John A. "Old-Regime Governors: Bureaucratic and Patrimonial Attributes." *Comparative Studies in Society and History: An International Quarterly* 14 (1972), 1–28.
——— "Sources of Administrative Behavior: Some Soviet and Western Europe Comparisons." *American Political Science Review* 59, no. 3 (1965), 643–55.
———. "Tsarist and Soviet Elite Administrators." *Slavic Review* 31, no. 1 (1972), 1–28.
Baginskii, I. M. "Rol' V. V. Kuibysheva v sovershenstvovanii deloproizvodstva NK RKI SSSR (1923–1926 gg.)." *Sovetskie arkhivy* 4 (1979), 13–14
Bailes, Kendall E. "The Politics of Technology: Stalin and Technocratic Thinking among Soviet Engineers." *American Historical Review* 79, no. 2 (1974), 445–469.
———. "Stalin and the Making of New Elite: A Comment." *Slavic Review* 39, no. 2 (1980), 286–289.

Bairoch, Paul. "Niveaux de developpement économique de 1810 à 1910." *Annales: Economies, Sociétés, civilisations* (1965), 1091–1116.
Barber, John. "The Establishment of Intellectual Orthodoxy in the USSR, 1928–1934." *Past and Present* 82 (1979), 141–64.
Benjamin, Roger, and John Kautsky. "Communism and Economic Development." *American Political Science Review* 62, no., 1 (1968), 110–23.
Bennett, Helju A. "Chiny, Ordena, and Officialdom." In Walter M. Pinter and Don Karl Rowney, eds., *Russian Officialdom: The Bureaucratization of Russian Society from the Seventeenth to the Twentieth Century*, 162–89. Chapel Hill: University of North Carolina Press, 1980.
——— "The Evolution of the Meanings of *Chin*: An Introduction to the Russian Institution of Rank Ordering and Niche Assignment from the Time of Peter the Great's Table of Ranks to the Bolshevik Revolution." *California Slavic Studies* 10, no. 1 (1977), 1–43.
Billington, James H. "Six Views of the Russian Revolution."*World Politics* 18 (1966), 452–73.
Blum, Jerome. "The European Village as a Community: Origins and Functions." *Agricultural History* 45 (1971), 157–78.
Carlisle, Robert B. "The Birth of Technocracy: Science, Society, and Saint-Simonians." *Journal of the History of Ideas* 35, no. 3 (1974), 445–64.
Casanova, José Vincente. "The Opus Dei Ethic and the Modernization of Spain." Ph.D. diss., New School for Social Research, New York, 1982.
Chechik, M. O. "Bor'ba komunisticheskoi Partii za ukreplenie i uluchshenie raboty sovetskogo gosudarstvennogo apparata v gody pervoi piatiletki (1928–1932 gg.)." *Vestnik Leningradskogo Universiteta: Seriia istorii, iazyka i literatury* 8, no. 2 (1958), 49–65.
Chikin, S. Ia. "Obrazovanie SSSR i razvitie narodnogo zdravookhraneniia." *Voprosy istorii KPSS* 10 (1972), 18–32.
Chizova, L. M. "K istorii pervykh mobilizatsii partiinykh rabotnikov v derevniu." *Sovetskie arkhivy* 2 (1968), 69–74.
——— "Vydvizhenie, 1921–1937 gg." *Voprosy istorii KPSS*, no. 9 (1973), 114–26.
Christian, David. "The Supervisory Function in Russian and Soviet History." *Slavic Review* 41, no. 1 (1982), 73–90.
Colton, Timothy J. "Military Councils and Military Politics in the Russian Civil War." *Canadian Slavonic Papers* 8, no. 1 (1976), 36–57.
Connor, Walter D. "Revolution, Modernization, and Communism." *Studies in Comparative Communism* 8, no. 4 (1975), 389–96.
Cox, Terry. "Class Analysis of the Russian Peasantry: The Research of Kritsman and His School." *Journal of Peasant Studies* 11, no. 2 (1984), 11–59.
Crews, Frederick. "In the Big House of Theory: Review of Quentin Skinner, 'The Return of Grand Theory in the Human Sciences.' " *New York Review of Books* 33, no.9 (1986), 36–42.
Daniels, Robert V. "The Secretariat and the Local Organizations in the Russian Communist Party, 1921–1923." *American Slavic and East European Review* 16, no. 1 (1957), 32–49.
Danilov, V. P., and T. I. Slavko. "O Putiakh issledovaniia dannykh nalogovykh svodok po sel'skomu khoziaistvu SSSR za 1924–1928 gg." *Istoriia SSSR* 5 (1972), 90–104.
Davies, James C. "The J-Curve of Rising and Declining Satisfactions as a Cause of Some Great Revolutions and a Contained Rebellion." In Hugh Davis Graham and

Ted Robert Gurr, eds., *The History of Violence in America*, 690–730. New York: Free Press, 1969.

———. "Toward a Theory of Revolution." *American Sociological Review* 27 (1962), 1–19.

Davis, Christopher. "Economic Problems of the Soviet Health Service: 1917–1930." *Soviet Studies* 35, no. 3 (1983), 343–61.

Derrida, Jacques. "Structure, Sign, and Play in the Discourse of the Human Sciences." In Richard Macksey and Eugenio Donato, eds., *The Structuralist Controversy: The Language of Criticism and the Sciences of Man*, 247–65. Baltimore, Md.,: Johns Hopkins University Press, 1970.

Deutsch, Karl W. "Social Mobilization and Political Development." *American Political Science Review* 55, no. 3 (1961), 493–514.

Dimock, Marshall E., "Bureaucracy Self-Examined." In Robert K. Merton, Alsa P. Gray, Barbara Hockey, and Hanan C. Selvin, eds., *Reader in Bureaucracy*, 397–406. New York: Free Press, 1952.

Drobizhev, V. Z., and E. I. Pivovar. "Large Data Sources and Methods of Their Analysis." In Don Karl Rowney, ed., *Soviet Quantitative History*, 141–68. Los Angeles: SAGE, 1983.

Eisenstadt, S. N. "The Social Framework and Conditions of Revolution." *Research in Social Movements, Conflicts, and Change* 1 (1978), 85–104.

Ekstein, Harry. "On the Etiology of Internal Wars." *History and Theory* 4 (1965), 133–63.

Ferro, Marc. "The Russian Soldier in 1917: Undisciplined, Patriotic, and Revolutionary." *Slavic Review* 30 (1971), 483–512.

Fitzpatrick, Sheila. "The Russian Revolution and Social Mobility: A Re-examination of the Question of Social Support for the Soviet Regime in the 1920s and 1930s." *Politics and Society* 13, no. 2 (1984), 119–41.

———. "The 'Soft' Line on Culture and Its Enemies: Soviet Cultural Policy, 1922–1927." *Slavic Review* 33, no. 2 (1974), 267–287.

———. "Stalin and the Making of a New Elite, 1928–1939." *Slavic Review* 38, no. 3 (1979), 377–402.

Freeze, Gregory L. "The Soslovie (Estate) Paradigm and Russian Social History." *American Historical Review* 91, no. 1 (1986), 11–36.

Getty, J. Arch. "Party and Purge in Smolensk, 1933–1937." *Slavic Review* 42, no. 1 (1983), 60–79.

Gill, Graeme J. "The Failure of Rural Policy in Russia, February–October, 1917." *Slavic Review* 37, no. 2 (1978), 241–58.

———. "The Mainsprings of Peasant Action in 1917." *Soviet Studies* 30, no. 1 (1978), 63–86.

Gimpel'son, E. G. "Rabochii klass v upravlenii sovetskim gosudarstvom: Noiabr' 1917–1920." *Voprosy Istorii* 11 (1981), 25–40.

Goldsmith, Raymond W. "The Economic Growth of Tsarist Russia, 1860–1913." *Economic Development and Cultural Change* 9, no. 3 (1961), 441–75.

Goldstone, Jack A. "Theories of Revolution: The Third Generation (Review Essay)." *World Politics* 32, no. 3 (1980), 425–53.

Gottschalk, Louis. "The Causes of Revolution." *American Journal of Sociology* 50, no. 1 (1944), 1–8.

Gouldner, Alvin W. "The Problem of Succession in Bureaucracy." In Alvin W. Gouldner, ed., *Studies in Leadership: Leadership and Democratic Action*, 644–59. New York: Wiley, 1950.

Grant, Stephen A. "Obshchina and Mir." *Slavic Review* 35, no. 2 (1976), 636–51.
Griffith, William E. "The Pitfalls of the Theory of Modernization."*Slavic Review* 33, no. 2 (1974), 246–52.
Groh, D. "Strukturgeschichte also 'Totale' Geschichte." *Vierteljahrschrift für Sozial- und Wirtschaftsgeschichte* 58, no. 3 (1971), 289–322.
Gunnell, John G. "The Technocratic Image and the Theory of Technocracy." *Technology and Culture* 23, no. 3 (1982), 392–416.
Gurr, Ted R. "A Causal Model of Civil Strife: A Comparative Analysis Using New Indices." *American Political Science Review* 62(1968), 1104–24.
——. "Psychological Factors in Civil Violence." *World Politics* 20 (1967), 245–278.
——. "The Revolution–Social Change Nexus: Some Old Theories and New Hypotheses." *Comparative Politics* 5 (1973), 359–92.
Guse, John Charles. "The Spirit of Plassenburg: Technology and Ideology in the Third Reich." Ph.D. diss., University of Nebraska, 1981.
Haimson, Leopold. "The Problem of Social Stability in Urban Russia, 1905–1917." *Slavic Review* 23, no. 4 (1964), 619–42.
Hamburg, G. M. "Portrait of an Elite: Russian Marshals of the Nobility, 1861–1917." *Slavic Review* 40, no. 4 (1981), 585–602.
——. "The Russian Nobility on the Eve of the 1905 Revolution." *Russian Review* 38 (1979), 323–338.
Hill, Alette O., and Boyd H. Hill, Jr. "Marc Bloch and Comparative History." *American Historical Review* 85, no. 4 (1980), 828–57.
Himmelstein, J. L., and M. S. Kimmel. "States and Social Revolutions: The Implications and Limits of Skocpol's Structural Model. (Review Essay)." *American Journal of Sociology* 86, no. 5 (1981), 1145–54.
Hobsbawm, E. J. "The Revival of Narrative: Some Comments." *Past and Present* 86 (1980), 3–8.
Holly, Douglas. "Learning and the Economy: Education under the Bolsheviks, 1917–1929." *History of Education* 11, no. 1 (1982), 35–43.
Hughes, Everett C. "Institutional Office and the Person." *American Journal of Sociology* 43 (1937), 404–13.
Husband, William B. "The Nationalization of the Textile Industry of Soviet Russia, 1917–1920: Industrial Administration." Ph.D. diss., Princeton University, 1984.
Iggers, Georg G. "Die 'Annales' und ihre Kritiker: Probleme moderner franzosischer Sozialgeschichte." *Historische Zeitschrift* 219, no. 3 (1974), 578–608.
Johnson, Robert Eugene. "Peasant Migration and the Russian Working Class: Moscow at the End of the Nineteenth Century." *Slavic Review* 35, no. 4 (1976), 652–664.
Jones, Gareth Stedma. "From Historical Sociology to Theoretical History." *British Journal of Sociology* 27 (1976), 295–305.
Judt, Tony. "Clown in Royal Purple: Social History and the Historians." *History Workshop Journal* 7 (1979), 66–94.
Kautsky, John H. "Revolutionary and Managerial Elites in Modernizing Regimes." *Comparative Politics* 14 (1969), 441–67.
Kanishcheva, N. I. "Sovremennaia zapadnogermanskaia burzhauznaia istoriografiia o predposylkakh Oktiabr'skoi Revoliutsii." *Istoriia SSSR* 4 (1983), 160–75.
Karabel, J. "Revolutionary Contradictions—Gramsci and the Problem of Intellectuals." *Journal of Politics and Society* 6, no. 2 (1976), 123–72.
Karanovich, G. "Etapy razvitiia mestnykh organov zdravookhraneniia." *Biulletin' NARKOMZDRAVa RSFSR* 20 (October 1927), 16–17).

Bibliography

Keep, John L. H. "The Agrarian Revolution of 1917–1918 in Soviet Historiography." *Russian Review* 36, no. 4 (1977), 405–23.
Kenez, Peter. "Liquidating Illiteracy in Revolutionary Russia." *Russian History* 9, no. 3 (1982), 173–86.
Kennedy, Charles H. "Technocrats and the Generalist Mystique: Physicians, Engineers, and the Administrative System of Pakistan." *Journal of Asian and African Studies* (Netherlands) 17, no. 2 (1982), 98–121.
Kerim-Markus, M. B. "Kadry NARKOMPROSa v pervyi god sovetskogo gosudarstva." *Istoricheskie zapiski Akademii Nauk SSSR* 101 (1978), 72–99.
Koenker, Diane. "Urbanization and Deurbanization in the Russian Revolution and Civil War." *Journal of Modern History* 57, no. 3 (1985), 424–50.
Kolychev, V. G. "Podgotovka kadrov politsostava krasnoi armii v gody grazhdanskoi voiny (1918–1920)." *Voprosy Istorii KPSS* 3 (1979), 66–76.
Korelin, A. P. "Rossiiskoe dvorianstvo i ego soslovnaia organizatsiia (1861–1904 gg.)" *Istoriia SSSR*, 5 (1971), 56–81.
Korelin, A. P., and S. V. Tiutiukin. "Sotsial'nye predposylki velikogo oktiabria." *Voprosy istorii* 10 (1970), 20–40.
Korolev, A. I. "V. I. Lenin i sozdanie sovetskogo gosudarstva." *Sovetskoe gosudarstvo i pravo* 3 (1970), 3–11.
Koval'chenko, Ivan D. "O burzhuaznom kharaktere krest'ianskogo khoziaistva evropeiskoi Rossii v kontse XIX-nachale XX veka (Po biudzhetnym dannym srednechernozemnykh gubernii)." *Istoriia SSR* 5 (1983), 50–81.
Koval'chenko, Ivan D., and N. Selunskaia. "Large Data Files and Quantitative Methods in the Study of Agrarian History." In Don Karl Rowney, ed., *Soviet Quantitative History*, 47–74. Los Angeles: SAGE, 1983.
Kramnick, Isaac. "Reflections on Revolution: Definition and Explanation in Recent Scholarship." *History and Theory* 11, no. 1 (1972), 26–63.
Krasovitskaia, T. Iu. "Dokumenty pervykh gosudarstvennikh organov po rukovodstvu prosveshcheniem natsional'nostei RSFSR (1917–22)." *Sovetskie arkhivy* 4 (1975), 77–80.
Kriashcheva, A. "K kharakteristike krest'ianskogo khoziaistva revoliutsionnogo vremeni." *Vestnik Statistiki* 5 (1920).
Kritsman, L. N. "Class Stratification of the Soviet Countryside." *Journal of Peasant Studies* (Great Britain) 11, no. 2 (1984), 85–143.
Krug, Peter F. "Russian Public Physicians and Revolution: The Pirogov Society, 1917–1920. Ph.D. diss., University of Wisconsin, 1979.
Kuleshov, V. I. "Deiatel'nost' KPSS po razvitiiu morskogo transporta v gody pervoi piatiletki." *Vestnik Leningradskogo Universiteta: seriia istorii, iazyka i literatury*, no. 3 (1979), 103–6.
Kuznetsov, N. L. "Partiinaia organizatsiia Petrograda v bor'ba za sokhranenie promyshlennosti goroda (Oktiabr' 1917–1920)." *Vestnik Leningradskogo Universiteta: Seriia istorii, iazyka i literatury*, no. 2 (1982), 11–16.
Laqueur, Walter. "Is There Now, or Has There Every Been Such a Thing as Totalitarianism?" *Commentary* October 1985, pp. 29–35.
Laverychev, V. Ia. "Iz istorii sozdaniia tsentral'nykh gosudarstvennykh organov upravleniia promyshlennost'iu v 1917–1918 g." *Vestnik Moskovskogo Universiteta* 22, no. 3 (1967), 3–16.
Levina, S. R., and N. I. Levchenko, "K istorii zavoevaniia tekhniko-ekonomicheskoi samostoiatel'nosti SSSR (1932–1937)." *Sovetskie arkhivy* 3 (1982), 26–30.
Lewin, Moshe. "The Disappearance of Planning in the Plan." *Slavic Review* 32, no. 2 (1973), 237–91.

———. "The Immediate Background of Soviet Collectivization." *Soviet Studies* 17 (1966), 162–97.
———. "More Than One Piece Is Missing in the Puzzle." *Slavic Review* 44, no. 2 (1985), 239–43.
———. "N. I. Boukharine: Ses idées sur la planification économique et leur actualité." *Cahiers du Monde Russe et Sovietique* 13, no. 4 (1972), 481–501.
Lieberman, Sima. "The Ideological Foundations of Western European Planning." *Journal of European Economic History* (Italy) 10, no. 2 (1981), 343–71.
Lieberstein, Samuel. "Technology, Work, and Sociology in the USSR: The NOT Movement." *Technology and Culture* 16, no. 1 (1975), 48–66.
Lievan, Dominic C. B. "Bureaucratic Liberalism in Late Imperial Russia: The Personality, Career, and Opinions of A. N. Kulomzin." *Slavonic and East European Review* 60, no. 3 (1982), 413–432.
———. "The Russian Civil Service under Nicholas II: Some Variations on the Bureaucratic Theme." *Jahrbücher für Geschichte Osteuropas* 29, no. 3 (1981), 366–402.
Likhachev, M. T. "Gosudarstvennye glavnye i osobye komitety vremennogo pravitel'stva." *Voprosy istorii* 2 (1979), 30–41.
———. "Izmeneniia v strukture MID Rossii v predoktiabr'skom periode." *Sovetskie arkhivy* 2 (1979), 47–49.
Lincoln, W. Bruce. "The Genesis of an 'Enlightened' Bureaucracy in Russia, 1825–1856." *Jahrbücher für Geschichte Osteuropas* 20, no. 3 (1972), 321–30.
Littlejohn, Gary. "The Agrarian Marxist Research in its Political Context: State Policy and the Development of the Soviet Rural Class Structure in the 1920s." *Journal of Peasant Studies* 11, no. 1 (1984), 61–84.
Luke, T. W., and Carl Boggs. "Soviet Subimperialism and the Crisis of Bureaucratic Centralism." *Studies in Comparative Communism* 15, no. 2 (1982), 95–124.
Lutchenko, A. I. "Rukovodstvo KPSS: Formirovaniem kadrov teknicheskoi intelligentsii (1926–1933 gg.)" *Voprosy istorii KPSS* 2 (1966), 29–42.
McAuley, Mary. "The Hunting of the Hierarchy: RSFSR OBKOM First Secretaries and the Central Committee." *Soviet Studies* 26, no. 4 (1974), 473–501.
Magerovsky, Eugene L. "The People's Commissariat for Foreign Affairs, 1917–1946." Ph. D. diss., Columbia University, 1975.
Maier, Charles S. "Between Taylorism and Technocracy: European Ideologies and the Vision of Industrial Productivity." *Journal of Contemporary History* 5, no. 2 (1970), 27–61.
Male, D. J. "The Village Community in the USSR, 1925–1930." *Soviet Studies* 14, no. 3 (1963), 225–48.
Malyshev, M. O. "Partiino-politicheskaia rabota v pervykh formirovaniiakh krasnoi armii, dislotsirovavshikhsia na territorii Petrograda i ego gubernii v 1918 g." *Vestnik Leningradskogo Universiteta: Seriia istorii, iazyka i literatury* 3 (1980), 5–11.
Mamulova, L. G. "Sotsial'nyi sostav uezdnykh zemskikh sobranii v 1865–1886 godakh." *Vestnik Moskovskogo Universiteta* 6 (1962), 32–47.
Manicas, Peter. "The Concept of Social Structure." *Journal for the Theory of Social Behaviour* 10, no. 2 (1980), 65–82.
———. "Review Essay: *States and Social Revolutions*." *History and Theory* 20, no. 2 (1981), 204–18.
Mann, M. "The Autonomous Power of the State—Its Origins, Mechanisms and Results." *Archives européennes de sociologie* 25, no. 2 (1984), 185–213.
Meehan-Waters, Brenda. "Social and Career Characteristics of the Administrative

Elite, 1689–1761." In Walter M. Pinter and Don Karl Rowney, eds., *Russian Officialdom: The Bureaucratization of Russian Society from the Seventeenth to the Twentieth Century*, 76–105. Chapel Hill: University of North Carolina Press, 1980.
Mendel, Arthur P. "Peasant and Worker on the Eve of the First World War." *Slavic Review* 24, no. 1 (1965), 23–33.
Merton, Robert K. "Bureaucratic Structure and Personality." *Social Forces* 18 (1945), 560–68.
——. "The Unanticipated Consequences of Purposive Social Action." *American Sociological Review* 1 (1936), 894–904.
Mommsen, Wolfgang J. "Die geschichtswissenschaft in der modernen Industriegesellschaft." *Vierteljahrshefte für Zeitgeschichte* 22, no. 1 (1974), 1–17.
Morozov, L. F. and V. P. Protnov. "Nachal'nyi etap v osusshchestvlenii Leninskikh idei o gosudarstvennom kontrole." *Voprosy istorii KPSS* 11 (1979), 33–44.
Mosse, Werner E. "Aspects of Tsarist Bureaucracy: Recruitment to the Imperial State Council, 1855–1914." *Slavonic and East European Review* 57, no. 2 (1979), 240–54.
——. "Aspects of Tsarist Bureaucracy: The State Council in the Late Nineteenth Century." *English Historical Review* 45, no. 375. (1980), 268–92.
——. "Makers of the Soviet Union." The *Slavonic and East European Review* 46, no. 106 (1968), 141–154.
Naumov, O. V., L. S. Petrosian, and A. K. Sokolov. "Kadry rukovoditelei spetsialistov i obsluzhivaiushchego personala promyshlennykh predpriatii po dannym professional'noi perepisi 1918 goda." *Istoriia SSSR* 6 (1981), 96–115.
Orlov, B. P. and R. I. Shniper. "Obzor opyta resheniia narodnokhoziastvennykh problem v sibiri." *Izvestiia Sibirskogo otdeleniia Akademii Nauk SSSR, seriia obshchestvennykh nauk*, no. 2 (1980), 11–19.
Orlov, V. S. "V. I. Lenin i nachalo deiatel'nosti Malogo SOVNARKOMa." *Voprosy istorii* 9 (1974), 210–13.
Orlovsky, Daniel T. "High Officials in the Ministry of Internal Affairs, 1855–1881." In Walter M. Pintner and Don Karl Rowney, eds., *Russian Officialdom: The Bureaucratization of Russian Society from the Seventeenth to the Twentieth Century*, 250–82. Chapel Hill: University of North Carolina Press, 1980.
Osipova, T. V. "Vserossiiskii soiuz zemel'nykh sobstvennikov (1917)." *Istoriia SSSR* 3 (1976), 115–29.
Panfilova, T. K. "Iz istorii kodifikatsii sovetskogo zemel'nogo zakonodatel'stva." *Sovetskie arkhivy* 5 (1976), 76–85.
Perrie, Maureen. "The Russian Peasant Movement of 1905–1907." *Past and Present* 57 (1972), 123–55.
Philp, Mark. "Foucault on Power: A Problem in Radical Translation?" *Political Theory* 11, no. 1 (1983), 29–52.
Pintner, Walter M. "Civil Officialdom and the Nobility in the 1850s." In Walter M. Pintner and Don Karl Rowney, eds., *Russian Officialdom: The Bureaucratization of Russian Society from the Seventeenth to the Twentieth Century*, 227–49. Chapel Hill: University of North Carolina Press, 1980.
——. "The Evolution of Civil Officialdom, 1755–1855. In Walter M. Pintner and Don Karl Rowney, eds., *Russian Officialdom: The Bureaucratization of Russian Society from the Seventeenth to the Twentieth Century*, 190–226. Chapel Hill: University of North Carolina Press, 1980.
——. "The Russian Higher Civil Service on the Eve of the Great Reforms." *Journal of Social History* (Spring 1975), 55–68.

———. "The Social Characteristics of the Early Nineteenth–Century Russian Bureaucracy." *Slavic Review* 29 (September 1970), 429–43.
Plavsic, Borivoj. "Seventeenth Century Chanceries and Their Staffs." In Walter M. Pintner and Don Karl Rowney, eds., *Russian Officialdom. The Bureaucratization of Russian Society from the Seventeenth to the Twentieth Century*, 19–45. Chapel Hill: University of North Carolina Press, 1980.
Presthus, Robert V. "Weberian vs. Welfare Bureaucracy in Traditional Society." *Administrative Science Quarterly* 6, no. 1 (1961), 1–24.
Protasov, L. G. "Klassovyi sostav soldat russkoi armii pered oktiabrem." *Istoriia SSSR* 1 (1977), 33–48.
Raeff, Marc. "Bureaucratic Phenomena of Imperial Russia." *American Historical Review* 84, no. 2 (1979), 399–411.
———. "The Russian Autocracy and Its Officials." *Harvard Slavic Studies* 4 (1957), 77–91.
Reimer, Reynold A. "The National School of Administration: Selection and Preparation of an Elite in Post-War France." Ph.D. diss., Johns Hopkins University, 1977.
Remington, Thomas. "Institution Building in Soviet Russia: The Case of 'State Kontrol.'" *Slavic Review* 41, no. 1 (1982), 91–103.
Remlinger, Gaston W. "Autocracy and the Factory Order in Early Russian Industrialization." *Journal of Economic History* 20, no. 1 (1960).
Richter, Melvin. "Tocqueville's Contributions to the Theory of Revolution." *Nomos* 8 (1967), 75–121.
Rigby, T. Harry. "The Birth of the Central Soviet Bureaucracy." *Politics (Australasian Political Studies Association Journal)* 7, no. 2 (1972), 121–34.
———. "Early Provincial Cliques and the Rise of Stalin." *Soviet Studies* 33, no. 1 (1981), 3–28.
Robbins, R. G. Jr. "Choosing the Russian Governors: The Professionalization of the Gubernatorial Corps." *Slavonic and East European Review* 58, no. 4 (1980), 541–60.
Romanovskii, N. V. "Istoriia velikogo oktiabria na stranitsakh zhurnala angliiskikh sovetologov." *Voprosy istorii* 1 (1975), 183–92.
Roos, J. P. "Theories of Planning and Democratic Planning Theory." *Government and Opposition* 9, no. 3 (1974), 331–44.
Rosenberg, D. "Budgets and Human Agency—Do the Men Have to Be Dragged Back In?" *Sociological Review* 33, no. 2 (1985), 193–220.
Rosenberg, William G. "The Democratization of Russia's Railroads in 1917." *American Historical Review* 86, no. 5 (1981), 983–1008.
———. "Russian Labor and Bolshevik Power after October." *Slavic Review* 44, no. 2 (1985), 213–38.
Rowney, Don Karl. "Higher Civil Servants in the Russian Ministry of Internal Affairs: Some Demographic and Career Characteristics, 1905–1916." *Slavic Review* 31, no. 1 (1972), 101–10.
———. "How History Beats the System: Disaggregative Characteristics of Open Political Systems in Revolution." In A. J. Neal, ed., *Violence in Human and Animal Societies*. Chicago: Nelson-Hall, 1976.
———. "Organizational Change and Social Adaptation: The Pre-Revolutionary Ministry of Internal Affairs." In Walter M. Pintner and Don Karl Rowney, eds., *Russian Officialdom: The Bureaucratization of Russian Society from the Seventeenth to the Twentieth Century*, 183–317. Chapel Hill: University of North Carolina Press, 1980.

――――. "Structure, Class, and Career: The Problem of Bureaucracy and Society in Russia, 1801–1917." *Social Science History* 6, no. 1 (1982), 87–110.
――――. "The Study of the Imperial Ministry of Internal Affairs in Light of Organization Theory." In Roger Kanet, ed., *The Behavioral Revolution and Communist Studies*. New York: Free Press, 1970.
Saccaro-Battisti, G. "Changing Metaphors of Political Structure." *Journal of the History of Ideas* 44, no. 1 (1983), 31–45.
Safraz'ian, N. L. "Stanovlenie sovetskoi sistemy vysshego obrazovaniia (1920–1927 gg.) istoriografiia: Problemy." *Vestnik Moskovskogo Universiteta, seriia 8: Istoriia* 1 (1981), 3–16.
Savel'ev, M. "Tsentral'nyi komitet vsesoiuznoi komunisticheskoi partii (bol'shevikov): Struktura TsK i ego organov." In *Bol'shaia sovetskaia entsiklopediia*, 16:547–60.
Savigear, P. "Some Political Consequences of Technocracy." *Journal of European Studies* 1, no. 2 (1971), 149–60.
Sentsov, A. A. "Detsentralizatsiia upravleniia v Rossii nakanune oktiabr'skoi revoliutsii." *Sovetskoe gosudarstvo i pravo* 7 (1981), 117–23.
Shanin, Teodor. "The Peasantry as a Political Factor." *Sociological Review* 14, no. 1 (1966), 204–18.
Sharlet, Robert. "The Soviet Union as a Developing Country: A Review Essay." *Journal of Developing Areas* 2, no. 2 (1968), 270–76.
Shtaerman, E. M. "On the Problem of Structural Analysis of History." *Soviet Studies in History* 7, no. 4 (1969), 38–55.
Siegelbaum, Lewis. "Production Collectives and Communes and the 'Imperatives' of Soviet Industrialization, 1929–1931." *Slavic Review* 45, no. 1 (1929), 65–84.
Simms, James Y. "The Crop Failure of 1891: Soil Exhaustion, Technological Backwardness, and Russia's 'Agrarian Crisis.'" *Slavic Review* 41, no. 2 (1982), 236–50.
Skocpol, Theda. "A Critical Review of Barrington Moore's *Social Origins of Dictatorship and Democracy*." *Politics and Society* 4 (1973), 1–34.
――――. "France, Russia, China: A Structural Analysis of Social Revolutions." *Comparative Studies in Society and History* 18 (1976), 175–210.
――――. "Old Regime Legacies and Communist Revolutions in Russia and China." *Social Forces* 55, no. 2 (1976–77), 284–315.
――――. "What Makes Peasants Revolutionary? A Review Article." *Comparative Politics* 14, no. 3 (1982), 351–75.
Skocpol, Theda, and Margaret Somers. "The Uses of Comparative History in Macrosocial Inquiry." *Comparative Studies in Society and History* 22, no. 2 (1980), 174–97.
Snegerev, M. "Velikaia Oktiabr'skaia sotsialisticheskaia revoliutsiia i raspredeleniia zemel' v 1917–1918 godakh." *Voprosy istorii* 11 (1947), 3–28.
Sofinova, R. L. and I. L. Nadezhdina, "Na Postu NARKOMA Putei Soobshcheniia." *Sovetskie arkhivy* 4 (1977), 43–52.
Sol'skii, D. I. "NOT i voprosy deloproizvodstva (1918–1924)." *Sovetskie arkhivy* 6 (1969), 47–52.
Sontag, John P. "Tsarist Debts and Tsarist Foreign Policy." *Slavic Review* 27, no. 4 (1968), 529–41.
Soskov, A. "Nekotorye voprosy upravleniia inzhenernymi voiskami v operatsiiakh tret'ego perioda velikoi otechestvennoi voiny." *Voenno-istoricheskii zhurnal* 23, no. 3 (1981), 28–34.
Sternheimer, Stephen P. "Administering Development and Development Admin-

istration: Organizational Conflict in the Tsarist Bureaucracy." Canadian–American Slavic Studies 9 (1975), 277–301.

———. "Administration for Development: The Emerging Bureaucratic Elite," 1920–1930." In Walter M. Pintner and Don Karl Rowney, eds., *Russian Officialdom. The Bureaucratization of Russian Society from the Seventeenth to the Twentieth Century*, 316–54. Chapel Hill: University of North Carolina Press, 1980.

Stone, Lawrence. "The Revival of Narrative: Reflections on a New Old History." *Past and Present* 85 (1979), 3–24.

———. "Theories of Revolution: The Third Generation." *World Politics* 18 (1966), 159–76.

Strizhkov, Iu. K., D. Tsiurupa, and A. D. Fokina. "Narodnye komissary Prodovol'stviia." *Istoriia SSSR* 3 (1971), 95–113.

Subbotin, D. T. "Ankety zheleznodorozhnykh iacheek RKP(b) kak istoricheskii istochnik (1919–1920 gody)." *Sovetskie arkhivy* 2 (1972), 76–79.

Suny, Ronald G. "Toward a Social History of the Russian Revolution." *American Historical Review* 88, no. 1 (1982), 31–52.

Therborn, Göran. "The Prospects of Labor and the Transformation of Advanced Capitalism." *New Left Review* 145 (1984), 5–38.

———. "Social Practice, Social Action, Social Magic." *Acta Sociologica* 16, no. 3 (1973), 157–74.

Tilly, Charles. "Does Modernization Breed Revolution?" *Comparative Politics* 5 (1972), 425–47.

Tilly, Louise A. "Problems in Social History: A Symposium." *Theory and Society* 9 (1980), p. 667–81.

Timasheff, N. S. "Social Mobility in Communist Society." *American Journal of Sociology* 50, no. 1 (1944), 9–21.

Torke, Hans-Joachim. "Das russische Beamtentum in der ersten Haelfte des 19. Jahrhunderts." *Forschungen zur osteuropaeischen Geschichte* 13 (1967), 7–345.

Trimberger, Ellen Kay. "A Theory of Elite Revolutions." *Studies in Comparative International Development* 7 (1972), 191–207.

Trofimov, L. I. "Pervye shagi sovetskoi diplomatii." *Novaia i noveishaia istoriia* 6 (1971), 37–52.

Uldricks, Teddy J. "The Crowd in the Russian Revolution: Towards Reassessing the Nature of Revolutionary Leadership." *Slavic Review* 36, no. 2 (1977), 187–204.

Vigne, Eric. "URSS: Les ingénieurs prennent le pouvoir." *Histoire* 26 (1980), 88–90.

Viola, Lynn. "Notes on the Background of Soviet Collectivization: Metal Worker Brigades in the Countryside, Autumn, 1929." *Soviet Studies* 36 (1984), 205–22.

Vladimirtsev, I. N., and A. S., Severin. "Voenno-politicheskii soiuz sovetskikh respublik (1917–1922 gody)." *Sovetskie arkhivy* 2 (1972), 3–22.

Von Laue, Theodore H. "Imperial Russia at the Turn of the Century: The Cultural Slope and the Revolution from Without." *Comparative Studies in Society and History* 3 (1960–61), 353–67.

———. "Russian Labor between Field and Factory, 1892–1903." *California Slavic Studies* 3 (1964), 33–65.

Weissman, Neil Bruce. "State, Estate, and Society in Tsarist Russia: The Question of Local Government, 1900–1908." Ph.D. diss., Princeton University, 1977.

Wheatcroft, S. G., R. W. Davies, and J. M. Cooper. "Soviet Industrialization Reconsidered: Some Preliminary Conclusions about Economic Development between 1926 and 1941." *Economic History Review*, 2nd ser. 39, no. 2 (1986), 264–94.

Wildman, Allan. "The February Revolution in the Russian Army." *Soviet Studies* 22, no. 1 (1970), 3–23.

Wright, Eric Olin. "To Control or to Smash Bureaucracy: Weber and Lenin on Politics, the State, and Bureaucracy." *Berkeley Journal of Sociology* 18–19 (1973–75), 70–108.

Zagorin, Perez. "Theories of Revolution in Contemporary Historiography." *Political Science Quarterly* 88 (1973), 23–52.

Zelnik, Reginald. "Russian Workers and the Revolutionary Movement (Essay Review)." *Journal of Social History* 6 (1971), 214–34.

Index

Abrams, Philip, 17
Agriculture, Commissariat of, 79, 80, 119, 120
Alexander I, ministerial reforms of, 20–24
Alexander II, 14
Alexander III, 201
Alexander Lyceum, 41, 60, 197
Anderson, Perry, 200
April Theses, 85
Armstrong, John A., 7–8, 38, 39, 42, 92, 183, 188, 196, 201, 203
Atkinson, Dorothy, 68
Azrael, Jeremy, 84

Bailes, Kendall E., 5, 6–7, 147, 173
Becker, Seymour, 30–31, 35, 36, 48, 57, 185
Bendix, Reinhard, 11, 30, 186, 200, 202, 206–7
Bennett, H. A., 97–98, 101
Bialer, Seweryn, 140, 202, 204
Bineman, Ia., and S. Kheinman, 132, 144, 147, 169
Blum, Jerome, 42–43, 47, 184–85
Bobrikova, V. N., 114
Bourgeois specialists, 88
Brezhnev, L. I., 29, 204
Bukharin, N. I., 95, 119, 121–22, 124, 126, 146, 194, 208

Bureaucracy: related to technocracy, 2–4; experts in, 2–4, 182–189; generalists in, 5, 186–87; defined, 9–13; organization type, 11; and revolution, 13–14; and modernization, 49–58; and control, 173–74, 176–82; and organizational adaptation, 190–91; and tsarist "*sluzhashchie*," 194–95

Careers and education, 33–34, 143–48, 163–65, 171, 202; and landholding, 35–38; and the nobility, 46–48, 57–61; and proletarianization, 148–59, 160; and bureaucratic politics, 207
Carr, E. H., 4–5, 165–66
Central Black Earth region, 73
Central Committee (of the All Union Communist Party), 141–43, 160
Cheka (*V.Ch.K*; RSFSR political police), 79
Chizhova, L. M., 164
Cholera epidemic, 82, 88
Colleges (Imperial *kollegii*), 22
Collegial organization (type), 10
Commerce, Imperial Ministry of, 21
Commissariat, *see individual commissariats*—e.g. Finances, Commissariat of
Commissariats, creation of, 28
Conquest, Robert, 193–94

233

Council of People's Commissars (SOVNARKOM), 80, 91, 128, 141, 142–43, 160, 186
Crozier, Michel, 179

Dagestan, 167
Davies, R. W., 4–5, 165
Davis, Christopher, 89
Debureaucratization, 126–27, 187, 205
Deutscher, Isaac, 122
Dimock, Marshall, 177
Dossiers, career (forlmuliarnye spiski), 23
Duma (legislature), 67
Dzerzhinskii, F. E., 112, 206

Education, Commissariat of, 82, 106, 107
Education, Imperial Ministry of, 21, 55
Eisenstadt, S. N., 199
Ekonomicheskaia gazeta, 189
Elites, status maintenance, 189–193
Eroshkin, N. P., 21, 66, 78, 79
Experts, 2–4, 5, 13–14; in Ministry of Ways of Communication, 39; in bureaucracy, defined, 182; and immobilization, 182–186; changing status after 1917, 186–189

Feedback, in complex systems, 178, 179
Finances, Commissariat of, 106, 107, 109, 110–11, 119
Finances, Imperial Ministry of, 21, 27, 55, 107
Fitzpatrick, Sheila, 116, 128, 150, 154, 163, 173, 202
Food Supply, Commissariat of, 79
Foreign Affairs, Commissariat of, 78
Foreign Affairs, Imperial Ministry of, 21, 99
Formuliarnye spiski, 23
Forrester, Jay W., 173, 178–179
Freiberg, N. G., 94, 103
Friedrich, Carl J., and Zbigniew K. Brzezinski, 180–81

Generalists, administrative, 5, 30, 38, 42–43, 63
Gerasimenko, G. A., 75
Gershenkron, Alexander, 121
Getlikh, L. F., 110
Getty, J. Arch, 151
Gill, Graeme, 70, 74
Gimpel'son, E. G., 107, 108
Governors, Imperial, 43–46, 58–62

Haimson, Leopold, 30
Hamburg, G. M., 30
Health, People's Commissariat for Preservation of, 79, 81–84, 91, 94, 95, 104, 105, 110; and public physicians, 82; creation of, 85–91; personnel in, 102–106, 119, 120.
Health, State Institute for Preservation of, 104
Holdovers, in revolutionary era, 109–20
Hough, Jerry F., 127, 202, 203

Inkeles, Alex, 131
Internal Affairs, Commissariat of, 42, 43, 55, 79, 82
Internal Affairs, Imperial Ministry of, 20, 21, 25–28, 95, 100, 101
Iroshnikov, M. P., 95, 107

Judge, Edward H., 180
Justice, Imperial Ministry of, 21, 55, 99

Karamzin, N. M., 1–2, 21–22, 26, 201
Karanovich, G., 86–87, 89
Kazakhstan, 167
Kazan province, 73
Khrushchev, N. S., 203, 207
Kirghizia, 167
Kleinmikhel, N. V., 60
Kniazhevich, N. A., 60
Kochubei, Viktor, 22
Komsomol (Communist Union of Youth), 133, 141
Korelin, A. P., 34, 35–36, 46
Kotel'nikov, K. G., and V. L. Meller, 70
Krokodil, 189
Krug, Peter, 96–97
Krupskaia, Nadezhda, 128

Labor, Commissariat of, 79
Lampert, Nicholas, 7, 195
Land: reforms in 1860s, affect on administration, 24–28; holdings of Imperial officials, 37; redistribution in 1917, 66; Decree of 1917, 72
Lenin, V. I., 1–3, 83, 84, 95, 119, 121, 122, 124, 125, 126, 127, 139, 141–42, 143, 171, 172, 186, 187, 193–94, 202
Lewin, Moshe, 165–66, 174, 187, 205, 209
Likhachev, M. T., 78
Local Economy, Main Administration for Affairs of, 49, 100, 101, 102, 104
Lorimer, Frank, 165, 166, 171

Index

Maine, Henry, 10
Male, D. J., 152, 165–66
Malevinskii, P. A., 111
Maliavskii, A. D., 72
Malinovskii, L. N., 99
Manning, Roberta Thompson, 14n.24, 30–31
Marine, Ministry of, 21
Marx, Karl, 126, 171
McNeill, William, 177, 184
Medical administration: Imperial, 98–102; Soviet, 102–6
Medical Council, Imperial, 100–102
Medvedev, Roy, 3–4, 9, 206
Ministerial government, creation of, 20–24
Ministers, Committee of, 22
Ministers, Council of, 41
Ministry, see individual ministries—e.g., Internal Affairs, Imperial Ministry of
Minsk province, 73
Modernization, 49
Mogilev province, 73
Moore, Barrington, Jr., 181

Nationalities, Commissariat of, 79
Nicholas I, 49, 55
Nicholas II, 4
Nobility: landed, 2, 14, 30–35; and bureaucratization, 38–43, 48–58, 61–62; and adaptation, 190–93; compared to other nobilities, 191–92
Nomenklatura, system, 120–21, 139, 143

Odom, William E., 188–89
Ogosudarstvlenie, 205
Organization, types of, 9–11
Orlovsky, D. T., 101, 147, 180
Owen, L. A., 74

Party (Russian Social Democratic or Communist), 3, 12, 83, 84, 141; and local administration, 72–77, 92; and upward mobility in, 131–39; and education requirements, 145; and proletarianization, 148–59
Pascal, Pierre, 124, 170
Patriarchal organization (type), 10
Peasants: and Ministry of Ways of Communication, 39, 69–77; activity in 1917, 68, 69; assemblies of (skhod), 69, 77; in Soviet administration, 130, 166–79; in provincial administration, 166–71; in Communist Party, 148–59

Penza province, 73
Personality, and interpretation in Russian history, 14–18
Physician-bureaucrats, 96, 106
Pintner, Walter M., 33–34, 43–45, 146
Police, Department of, 79
Posts and Telegraphs, Commissariat of, 106, 112–14
Posts and Telegraphs, Imperial Main Administration for, 112–14
Potulov, B. M., 83
Proletarianization, 128–29, 165–67; and Communist Party membership, 148–59; and white-collar status, 159–65; and political control, 171–74
Provinces, 68, 72–77, 76, 80, 145, 166–71; bureaucratization of, 48–57
Provisional Government: and provincial administration, 65, 67, 68–69, 205; and central administration, 77–81, 127

Raeff, Marc, 127
Rashin, A. G., 74
Red Army, 81
Red managers, 128
Red specialists, 87
Rein, G. E., 82, 91, 98–101, 106
Reports, Imperial (Vsepodanneishie doklady), 21
Revolution, xii, 1, 4–5, 8, 13–14, 28, 31, 45, 68, 87, 122, 131, 146, 178
Rieber, Alfred, 80, 184, 191
Rigby, T. H., 78, 87, 97, 128, 130, 139, 141–42, 143, 150, 152

Saltykov, Mikhail, 17–18
Sanovniki (top Imperial officials), 37
Saratov province, 73
Sectors, in Imperial administration, 38, 43–46, 57–58
Selznik, Philip, 181
Semashko, N. A., 87, 105
Seniority: in Imperial civil service, 11, 33, 101–2, 110; related to turnover in USSR, 148–59
Service, statute on, 22–23
Shakhty trial, 128
Shanin, Teodor, 165–66
Silberman, Bernard, 10
Simbirsk province, 73
Simirenko, Alex, 188
Skhod, 69, 77
Skocpol, Theda, 13, 31–32, 34, 179–80, 185, 192

Sluzhashchie (white-collar employees), 95, 130, 156, 159, 164, 167, 194–95, 197
Smirnov, P. A., 110
Smitten, E., 154–57, 162, 166
Smolensk province, 73
Social class: and Imperial bureaucracy, 30–35, 38–57; categories in 1920s, 130–31; emerging differences in 1920s, 138–39; differences and political leadership, 141–43; and upward mobility, 193–95; and turnover after 1917, 197–98
Social Welfare, Commissariat of, 79
Soloviev, Z. P., 88
Sorokin, Pitirim A., 178, 194
Soviets, executive committees (*ispolnitel'nye komitety* or ISPOLKOMy), 76, 80, 145; formation of, 67, 72–77, 88, 131
Stalin, J. V., 4, 14, 124, 139, 140, 143, 152, 165, 173, 174, 181, 187, 198, 207–9
Stalinism, transition to in 1920s, 139–40, 181, 194–95, 209
Starr, S. Frederick, 48–49
Statistical Administration, Central, 26, 27, 73, 119, 120
Statistical Department, All Union Communist Party, 149
Statization, 205
Sternheimer, Stephen P., 31, 131, 154, 164
Stone, Lawrence and Jeanne F., 37–38, 190–91
Structure and analysis, 15–17; and explanation, 17, 201; and social characteristics, 32; transformation of, 66
Sub-elites, upward mobility of, 109–18
Supreme Council of the National Economy (VSNKh), 106–9, 120, 186, 206

Talmon, J. L., 200
Tarassevitch, L. A., 87
Taylor, F. W., 128
Technical administration, introduction in Russia, 24–28
Technocracy defined, 5–9
Technocrats, 3, 6, 22, 24, 30, 62–64, 81–85, 95, 112; and development of the Soviet technostructure, 207–9
Tocqueville, Alexis de, 177–78, 192
Trade and Industry, Imperial Ministry of, 41
Trade, Commissariat of, 79
Trimberger, Ellen Kay, 13, 34
Troitskii, S. M., 35
Trotsky, L. D., 15, 119, 122, 126, 140, 142, 146, 151, 170, 171, 173, 174, 194, 208
Turnover of officials between 1890 and 1939, 195–205; in 1920s, 118–23, 148–59
Typhoid epidemic, 82
Typhus epidemic, 82, 88

Ukraine, 73, 167
Urbanization, and Imperial bureaucracy, 48–58
Ustav o sluzhbe, 22–23
Uzbekistan, 167

Vasiaev, V. I., 114, 116, 144, 147
Veterinary Administration, Imperial, 26
Vice-governors, 45–46
Vitebsk province, 73
Volga region, 73, 75
Von Laue, Theodore H., 12, 39–41
Vsepodanneishie doklady, 21
VSNKh. See Supreme Council of the National Economy

War, Ministry of, 21
Ways of Communication, Commissariat of, 106, 109
Ways of Communication, Imperial Main Administration of, 27, 55, 183
Ways of Communication, Imperial Ministry of, 27, 39–41, 96, 111–12, 183
Weber, Max, 11
Witte, S. Iu., 39, 41
Women in medical administration, 102
Wortman, Richard, 180

Yaney, George, 25, 34

Zaionchkovskii, P. A., 35–35, 46, 47
Zemstvos (land assemblies), 25, 68, 69, 82
Zubovskii, M. I., 61

Library of Congress Cataloging-in-Publication Data
Rowney, Don Karl, 1936–
   Transition to technocracy.
   (Studies in Soviet history and society)
   Bibliograhy: p.
   Includes index.
   1. Soviet Union—Politics and government—1917–
2. Technocracy.   I. Title.   II. Series: Studies in
Soviet history and society (Ithaca, N.Y.)
JN6531.R68   1989   306'.24'0947   88-47925
ISBN 0-8014-2183-7 (alk. paper)